BLUE
ISLAND

Other books by Jean Raspail

Who Will Remember the People . . .

The Camp of the Saints

Welcome Honorable Visitors

BLUE
ISLAND

A NOVEL BY
JEAN RASPAIL

Translated by Jeremy Leggatt

Mercury House, Incorporated
San Francisco

Published in the United States by
Mercury House
San Francisco, California

Distributed to the trade by
Consortium Book Sales & Distribution, Inc.
St. Paul, Minnesota

Printed on acid-free paper
Manufactured in the United States of America

Library of Congress Cataloging-in-Publication Data

Raspail, Jean.
  [Île bleue. English]
  Blue Island : a novel / by Jean Raspail ; translated by Jeremy Leggatt.
    p.   cm.
  Translation of: L'Île bleue.
  ISBN 0–916515–99–0 : $17.95
  1. World War, 1939–1945 — France — Fiction.   I. Title.
PQ2635.A379I413   1991
843'.914 — dc20   90–49381
                                                                      CIP

To Maïté

A MONTH IN bed followed by a slow convalescence, with barely the strength to move about within the four walls of my bedroom, and the equivocal happiness of total detachment . . . Visitors' voices reached me as though filtered through thick fog, their words awakening not the slightest interest in me. Books and newspapers lay in heaps on the table where they had fallen from my hands as soon as I tried to read them. My radio was silent. So I have no need to wonder whether outside factors, triggered by people or ideas, influenced the writing of this account forty-seven years after the event or influenced my decision to undertake it as soon as my health returned. They did not. For me, moreover, this story exists outside time. There is no other way to accept the eerie light that falls, as if from a dying star, from the blazing confrontation between Bertrand Carré and Frantz.

I was apparently very ill indeed. A raging and mysterious fever kept me totally bedridden for two weeks. My memories of that time are fraught with wonder. First, voices — clear, sharply defined, speaking French or German — saying some things that had actually been said and others that perhaps had not, but falling into a natural sequence during this astounding resurgence of memory: Bertrand Carré's imperious tones, the voices of Lieutenant Frantz von Pikkendorff and his black-uniformed tank crews, Zazanne's voice, heavy with animal sensuality, Maïté's aristocratic delivery, Pierrot's voice, Zigomar's, and my own, my thirteen-year-old voice — children's voices, adolescent voices, with the exception of the German voices, which were the voices of twenty-year-olds. And then there were other voices in the surrounding landscape, the voices of a country falling apart. I believe I heard them all much more clearly than I had forty-seven years earlier: the grinding of tank treads on the dirt road by the bridge over the Mulsanne, the water's song at the little dam just upstream of Blue Island, the whole spectrum of sounds and voices, horses stamping, cart wheels creaking, horns and car engines, as well as the protests and invective of a nation in headlong flight down the road from Loches to La Roche-Posay, and above all the extraordinary totality of the silence that preceded the appearance of the three tanks under von Pikkendorff's command, as though the ordainer of our destinies had erased everything round about, the better to focus his attention on us. Prostrate, burning with fever, eyes closed, my head swimming with visions, at once conscious and unconscious, I relived each episode down to its last detail, not once but a hundred times, like a film ceaselessly replayed, with occasional returns to the surface to drink the glass of water offered to me, to mumble replies to the questions of the doctor or my wife, and then to dive passionately back into the maelstrom of my memories. At certain points my soul was flooded with light — as when, for example, Bertrand Carré's dazzling beauty shone forth like a revelation, a masterpiece of adolescent grace, of natural distinction, of authority of gesture and gaze, brown of skin, black of hair, the elegance of a "pharaoh's son," as my aunt Melly put it, her words returning to me in my fever long years after I had forgotten them, just as I had long since erased all admiration for Bertrand from my mind.

As the feverish saga drew to a close, as strength and the inclination to make notes returned, I realized that *this time* I had forgotten nothing. This was enormously surprising to me, for I have little memory of my childhood or my adolescence, hardly even of early adulthood, and anything that might bring them back I destroy. Correspondence, manuscripts, photos, personal mementos, I throw them all away, I shred, I burn, without a thought, without a regret. Like a coughing fit, the need to destroy them periodically takes hold of me. Nothing escapes it. There is no familiar landscape behind me, merely vague memories like shapeless overgrown ruins. Doors close one after the other at my heels as I advance through life. When I turn back there is nothing there; when I die, nothing will have happened. There is probably some reason for this instinctive yet deliberate flight from memory. I once wrote, "I cannot understand what it is that childhood left in each of us — or even that it left anything at all. Yet childhood stubbornly forces us to admit to ourselves what we really are. . . ." It is probably there that we should seek the answers. Cruelty, cowardice, promises made to ourselves and broken, convenient disguises, borrowed attitudes — I must often have behaved in a manner inconsistent with my own pride, and since I was determined to keep that sense of pride intact in order to preserve a flattering image of myself, I simply forgot.

It is thus, in all likelihood, that I had forgotten Bertrand.

I HAD ALWAYS lived in Paris, where my father was a senior civil servant—and apparently a very successful one—attached to a succession of key government departments. Taking our cue from him, we invariably pronounced the names of the various ministers he had served with a mixture of respect and disgust. In those days of crisis, then of war, they were the masters at whose bidding my father stood, day and night; it hurt me to hear him, on the living room telephone, complying with the orders of politicians he privately despised, serving them to the best of his ability. The last of them was called Chabannais, post office minister I think he was, or perhaps industry, in Prime Minister Paul Reynaud's war cabinet. My father said his name sounded like a brothel, and suited him perfectly, an observation quite beyond me at the time.* We will cross paths with this character again, on June 11,

*Chabannais was the name of a famous brothel in Belle Époque France. — TRANS.

1940, at Château de La Celle, the home of my aunt Melly Lavallée in south Touraine, during the government's pathetic flight from Paris to Bordeaux, an ignominious hundred-hour scramble for which France is still paying the bill. On that day I decided to follow Bertrand Carré to Blue Island, and Chabannais was a powerful factor in my decision . . .

My father never took vacations. Ever since the Popular Front government had decreed that annual vacations were every Frenchman's right, he — a man who lived only for duty — had ostentatiously refused to take advantage of them. But the truth was that he loathed holidays, he detested the countryside and the sea, hunting, sport, and travel, and was bored nearly to death if he stayed more than a day away from the corridors of government and his book-filled Passy apartment, his whole world. Extremely close, he and my mother never left Paris. I was dispatched "for my health" to one of my aunts in Touraine, between Loches and La Roche-Posay. Within a radius of five miles I could count at least a half dozen of them at various removes of kinship: Aunt Octavie, Aunt Melly, Aunt Germaine. There, nestled in lovely little valleys, they lived in rustic manor houses with pretensions to château status; and there, hordes of cousins and friends congregated every summer, summers that lasted until the first of October, the date school reopened in those far-off days. It was another age. This book is not an attempt at nostalgia; yet (even though they would have been inconceivable without the reprieve afforded by the Munich agreements) those last summers of peace were truly blessed: all innocent pleasure, all high-spirited style, a style shared with our elders and adapted to the pleasures of the young — bicycle rides, white pleated skirts for the girls, long shorts for the boys, picnics, bathing in the Mulsanne, tennis, croquet, bread and bars of chocolate for tea (and the unvarying ritual of paying our respects to the aunts who took their tea in the arbor and whom we always addressed formally as *vous* and *ma tante*). Impeccably French (if the expression still has any meaning), the aunts seemed interchangeable. But from the faces that swirled through my nights of fever I most often recall that of my aunt Melly, my hostess at Château de La Celle. Aunt

Melly and her surprisingly youthful voice. Her way of scrutinizing us from head to toe, severely but with a friendly and prodigiously inquisitive gleam in her eye, before uttering the few words that released us to our games and escapades, the way she scrutinized us the day Bertrand Carré — the boy I had brought to introduce to her — stood before her with easy insolent grace.

"You're going to break many hearts, young man. Where are you from? What storms will you be bringing us?"

Bertrand merely answered, "Madam," bowing slightly and keeping his blue eyes fixed on hers.

Nothing escaped Aunt Melly: the black rebellious hair, the long eyelashes, the small straight insolent nose, the mouth with chiseled lips half open over gleaming teeth, the long brown legs, the delicate hands, and that oval face so many French adolescents possess, faces curiously feminine in appearance — but faces in which it would be foolish to read a lack of masculinity. The examination over, my aunt let out a little sigh of satisfaction and dismissed us with a wave of her hand.

"Don't get into too much trouble, you two."

It was the summer of '39. Bertrand Carré came to us from Versailles, he too to stay with an aunt, one who answered to the Christian name Sophie, two miles from La Celle on the other side of the Mulsanne. She quite obviously adored him, and gave him a much freer rein than was then considered advisable for a boy of his age, thirteen, just a few weeks' difference from me. His father, an air force squadron leader, had recently taken command of Lang-Son airfield in Tonkin, facing the Japanese across the Chinese frontier. Of his mother he never spoke. I had seen a photograph of his father, whom he did not resemble in the least. He must have taken after his mother. To complete this description I should add that he wore a bracelet, not a chain, which would have looked tawdry, but a true piece of jewelry, gold with bright enamelwork, fitting snugly around his left wrist like a strange scintillating cuff, mysteriously useless. Flashy and ostentatious, the thing nevertheless had a certain elegance; it was so well suited to the extraordinary personality of the boy who wore it so naturally that my aunt Melly and all the other aunts decided independently of one another (and despite the sarcastic protests of uncles and fat-headed older cousins) that it was in perfect taste simply because Bertrand was Bertrand — enough said! That bracelet hypnotized girls. As for other boys of our age who might have been tempted to sneer, one look from Bertrand Carré was enough to deter them.

Why have I spoken at such length of the bracelet? Perhaps because it is a symbol. What do we know about children's ideas? Don't children instinctively sense that the seemingly ridiculous is real? That is why they play. Play is the only escape from this perception, until the day they understand (provided they have remained pure of spirit) that convictions amount to nothing much more than attitudes and that the only way of remaining true to them is to play — and if need be to play for one's life, and to lose it. For all that, there were contradictions in Bertrand's character I won't try to explain. Later on, on Blue Island, he held the bracelet to the sunlight, making it sparkle like some psyche-delic light, and told me, "It belonged to my mother, and it's one hundred percent junk! Right out of a brothel . . ." That very evening I rushed to the dictionary, which from B (Brothel: House of prostitution) sent me to P, where prostitutes turned out to be women "who practice prostitution, giving themselves to men for money," which, although I was unclear as to the technical details of the transaction, made me blush to the roots of my hair and cast anxious looks at the half-open library door that Aunt Melly could have come through, dishonoring me forever with one glance at the dictionary. That's how it was for us boys, for us well-brought-up boys, in the summer of '39 . . .

I remember the start of the hunting season, in late August 1939. It was at Beausoleil, my uncle Léonce Bonnadieu's estate. Aunt Germaine's husband, he was inseparable from his hounds and was a renowned organizer of shoots between the Creuse and the Indre rivers. As we waited for the starting signal, Bertrand and some fifteen other boys, myself among them, sat watching from the steps of the main stairway; we had been hired for the day as beaters, armed with stationmaster-style red flags and paid a whole franc, an enormous goose-liver sandwich, and a bottle of apple cider. As each huntsman's name was called, Bertrand's eye would linger provocatively on his prancing, self-conscious elder, conceited, potbellied, hung about with cartridge belt and game bag, rigged out like President Kruger's Boers, and wearing a green hat decorated with pheasant feathers. Ridiculous these men assuredly were, at least to my uprooted city-boy's eyes, but no more so than today's fake commandos in jungle-green gear stalking tame pheasant in Sologne. As the dogs wagged frantic tails, the huntsmen milled around Emperor Léonce like Napoléon's generals on the morning of the Battle of Austerlitz, draining glasses of chilled Vouvray served by mustachioed gamekeepers buttoned to the chin. Bertrand Carré gave me a nudge.

"They're off to war," he said. "Raring to go. Cast-iron morale."

People talked a lot about war that year. Or at least about the one we had just gallantly avoided. For all these huntsmen had been mobilized the previous summer, only to return in the aftermath of Munich to their estates. They were just in time for the sacrosanct opening of partridge season: after the shoot, row upon row of dainty little feathered corpses would be lined up on the quiet gray cobblestones of the courtyard at Beausoleil, a glowing brown and gold still life in the classical manner, a bag that swelled our uncles and older cousins with pride even as Hitler flaunted his own trophies — the mortal remains of Czechoslovakia.

"And there!" said Bertrand. "Look at the widows!"

"The widows?"

A short distance from the Boer commandos the aunts and female cousins were more discreetly assembled around the convertible Citroen C4's in which they would follow the hunters with picnic baskets and bottles. Many of them knew how to drive and sported leather gloves for the occasion. Women didn't hunt in those days — any more, for that matter, than they voted. But in one case it was a question of law, and, in the other, it was how the menfolk liked it. If women hunt in Touraine today, it is because the men lost the war. I saw Bertrand exchange a knowing look with his aunt Sophie. She was watching her husband, my uncle Armand Majorel, with ironic commiseration as he mimed the start of last year's shooting by pointing his gun skyward, training it on an imaginary flight of partridge. The poor man . . . he was fat and solemn, long-winded and harmless. But ten months later, June 17, 1940, standing in his pigeon-tower, he redeemed himself by hurling insults at the Stuka dive-bombers that were machine-gunning the Indre bridges—leaving him (to his surprise and mortification) unscathed.

"Why 'widows'?" I asked.

"Just look at them!" Bertrand said. He waved at the knot of hunters. "Look at those caricatures! I bet the women would like to see them dead, laid out gloriously among the rabbits and the partridges with their dogs howling."

"Funny idea," I said.

"You call that funny? Hunting's like war, it should be a serious and dangerous business, with deaths on both sides. And for women there is only one role that can be played with dignity, that of the widow!"

Bertrand Carré was only thirteen years old in the summer of 1939. So was I. The other beaters laughed surreptitiously, except for Pierrot, son of one of Uncle Léonce's farmers, and Adhémar Durand, whom we called Zigomar, a shy, shortsighted beanpole attached to a gigantic nose as ridiculous and poorly suited to his face as his crusader's first name was to his dirt-common last name. Like me, they had already chosen sides.

A girl came over to join us, Maïté, the daughter of my aunt Octavie de Réfort. She too was thirteen, long and lean, almost without curves, with glorious blonde hair that she played with like a grown woman and gray eyes that came alive only at the sound of her own voice. Her skin was very white, her little breasts only just budding, with tips of a pearly pink, and she had a neat and perfect triangle of blonde shadow below her belly; I knew this because I had seen her naked several times, in Aunt Sophie's abandoned omnibus and later on Blue Island. She appeared to me thus so often during my long nights of fever that I can now evoke her image without difficulty, exactly as a man would, although I was then a child and loved her chastely. For I did love her, as a youngster could love in those days, without daring to take her hand, or to brush against her, or even to admit it to her, which would have been permissible according to the code we played by. She would have sent me packing. Her god was Bertrand.

As for me, I had to content myself with declaring my ardor to Suzanne Charpentier, daughter of the butcher, a good man who had taken to drink since losing his wife and who let his daughter run wild in our company . . . We called her Zazanne, and that says it all. Sweet, empty-headed, scatterbrained, warm and willing, she was bestowed on me with feudal and somewhat disdainful largesse by Bertrand, to whom she was of course slavishly attached. She had a speech defect, a guttural quality to her voice, so that as soon as she opened her mouth, which was fleshy and vulgar, like her chunky, wriggly little bottom, our thoughts turned to things both mysterious and smutty, and it is that voice I still hear. Today, having lived once again through visions of Zazanne, naked and heavily scented with cheap lavender water, and having in the half-delirium of fever found a shameful satisfaction in such memories, I realize that the prize allotted me by Bertrand was truly the small change of love. Even at thirteen. Already at thirteen. The easy, commonplace girl—she was the one for me. He had so ordained it. Juliet—the miraculous, golden-haired, delicate Maïté—was for Romeo, in the person of Bertrand Carré, master of our games but not of our destinies, because we were not worthy of him. And this too must have been one of my secret reasons for burying Bertrand Carré's grave on Blue Island beneath shovelfuls of oblivion.

Bertrand said to Maïté, "Do you want to be my widow?" The other kids rolled on the ground, laughing with senile incomprehension (for it is possible to be senile at any age). Pierrot and Zigomar watched their sovereign in adoration.

Like a young filly, Maïté tossed her blonde hair. "I am your widow, and you know it."

There is nothing more beautiful to me or more hateful than that particular scene, since I was excluded from it from the outset. As I lay alone on my damp sheets, I realized that it had marked my whole life.

Emperor Léonce, my uncle, blew on a small silver whistle that hung around his neck, sending a disciplined French army trotting out to war: gamekeepers, beaters, and uncles and prancing cousins who hurled themselves across fields and hedgerows to show the partridge and hare what France was made of. We waved our flags energetically, the flapping flushing out the game. We had to yell "Partridge, partridge!" or "Rabbit, rabbit!" if we saw one, to let the gentlemen on horseback know at once whether their honor was at stake aloft or on the ground. It was a memorable orgy of shooting.

Bertrand walked with teeth clenched, as though marching on an enemy. He had thrown away his flag, saying to me, "This is disgusting!" On several occasions he strayed dangerously from the line of beaters, and the bullets whistled about his ears as he walked without lowering his head.

"Little bastard!" said Raymond Bonnadieu, his arm raised as if to strike. He was a big, redheaded cousin who hated Bertrand. "I missed a whole covey" — of partridge — "because of you. And don't you realize that at that distance I had a fifty-fifty chance of killing you?"

"My my!" was Bertrand's reply. Then, indicating the landscape and the skirmish line of Boer marksmen moving into the next field, he added these strange words: "On the field of honor, you always die for nothing. Some know it, some don't."

"One of these days your conceit will kill you," said the cousin.

I should add that this cousin, an infantry lieutenant, died gallantly and in vain at Arras and left a most suitable widow who most suitably consoled herself, just after a hunt, once peace returned . . .

That very evening, at Beausoleil, after the ritual display of the bag (so big it was almost metaphysical), as the huntsmen sat in pheasant-feather hats partaking of hare pâté and Bordeaux, the radio announced that the two giants of the day had signed the

Soviet-German Pact. Then there was a mix-up in the station's music programming — Radio Paris, I think it was — unless it was on purpose, an extremely French way of bolstering the national morale. The voice of Maurice Chevalier, unctuously proletarian, relentlessly working-class, wheedling, sang, "Have some fun, ev'ryone, life goes by like a dre-e-am. Have some fun, ev'ryone, drink life to the dregs and *whee!*" A hit in those days. Silent before their rows of dead partridge and their plates of pâté, the huntsmen chewed over both news and song. The last stragglers made their entrance, their guns pointed toward the ground as though for an official funeral. The radio concluded: "And we're not here to think about care, we're not in this life to think about strife. Have some fun, ev'ryone," etc.

"Ah, no!" said Bertrand's imperious voice.

And the radio fell silent. Bertrand had clipped Momo's wings (in those days that was what Chevalier's fans called him). This happened again a few days later when the Germans crushed Poland ("Paris will always be Paris . . . We'll be hanging out the washing on the Siegfried line . . ."), but this no longer has to do with my story . . .

It must have been five o'clock. Crushed, long-faced, the huntsmen returned to their pretty little châteaux to break out their officers' uniforms, in mothballs ever since Munich. Bertrand assembled his clan with a look.

"The old bus, in half an hour!" he said.

WE CHILDREN WERE quite free in those days, about as free in fact as children of the same age today, but not for the same reasons. Since we were well brought up it was assumed that we would not do anything too terribly wrong, which was more or less true. The only limits imposed on our wanderings were mealtimes, set to the minute, with hands washed and hair combed — unless a picnic had been authorized — and the obligation to announce where we were going and to receive permission to leave, which was rarely refused. It was enough for me to tell my aunt Melly, "I'm leaving for La Jouvenière," to wait for her reply, and then jump on my bicycle to join my friends, who enjoyed the same easy rules.

La Jouvenière, the seat of Aunt Sophie Majorel, Bertrand's aunt, lay three miles from La Celle, where I lived with my aunt Melly Lavallée. About the same distance away were La Guichardière, the home of Aunt Octavie de Réfort, Maïté's mother; Beausoleil, home of Aunt Germaine Bonnadieu; and La Cornet-

13

terie, home of the Durands, where their only son, Zigomar, stayed. Zazanne lived with her butcher father in the village of Petit-Bossay, close to the center of our little universe. As for Pierrot, the farm run by his father bordered on Beausoleil. We hardly ever stopped there except to plunder the larder for goose liver. All this in an area not much bigger than a pocket handkerchief, but with hills and dales, small forests, large woods, hedged fields, an old mill, a ruined keep, a lake, and the winding Mulsanne with its bridge close to Blue Island, all of which inflated our tiny and really somewhat godforsaken corner of Touraine to the dimensions of a kingdom. The trails, known as rights-of-way (a charming expression now rapidly falling into disuse), were decently maintained gravel roads. Only Route 41, from Loches to La Roche-Posay, boasted an asphalt coat. It lay three miles to the east of the village of Petit-Bossay, below La Guichardière, but we were forbidden to cross it because of the risk of being run over. Everyone called it "the main road." As for signs of progress (apart from electricity, which failed on stormy evenings), there was just one manual gas pump in front of the village grocer's and just five crank-operated telephones connected to Petit-Bossay post office, Number 1 being my aunt Melly, who found in this primacy a boundless source of sarcasm and satisfaction.

Most often La Jouvenière was our meeting place. The Majorel family had seen better times. The house, a vast pile with a mansard roof, had a sturdy-looking façade, but the rest was on the road to ruin. The shutters needed repainting; the door frames were dilapidated; the decapitated pigeon-tower had been gnawed away by the same wind that had shaken its roof tiles loose; the garden gate and garden furniture were rusty; the gardens had been more or less abandoned. To us, it was an adventure-filled paradise. Children settle more readily into premises eroded by time than into dwellings as lovingly tended as my aunt Melly's, for example, which I fled at every opportunity. Our special spot, our hideout, the theater of all our dreams, was the carcass of an ancient black horse-drawn omnibus abandoned in an old carriage house at the far end of La Jouvenière's grounds.

As it loomed again in my fevered dreams it seemed enormous, a mountain perched on four wheels as tall as houses and cradled on leather and iron springs that looked like spiders' legs, with an Ali Baba trunk in the rear where twenty outlaws fleeing the tyrant's henchmen could have hidden, and (rearing as high as a ship's bridge except that it was closer to the bow) a coachman's seat high above the gallows-shaped harnessing pole, complete with whip mounted upright in its socket like a medieval banner. Children create a virgin forest out of the smallest copse, an underground Templar earthwork out of a cellar corridor, a stark and inaccessible frontier out of a hedge at the bottom of a garden. The fever had exaggerated my childhood memories tenfold, but even so this "bus" was a landau of respectable size that in the Majorel family's palmier days had ferried passengers between the Châtillon-sur-Indre train station and La Jouvenière—hence its utilitarian name. Forgotten for twenty years, it was in perfect condition, sheltered behind the enormous entrance of the carriage house, locked and missing its key. We got into the building by way of a high window furnished with a ladder. At close of day, shafts of sunlight would slant down through the gaps made by displaced roof slates, making the omnibus glow like a reliquary under stained glass. The window cranks still turned, the hinged running boards could still be raised and lowered, and closing the carriage doors produced a muted thump that seemed to reach us from another age. The lamps still held short, fat candles, and the black padding of the seats, designed for eight travelers of ample size sitting four by four, facing one another with their legs more or less outstretched, still boasted its leather buttons. The six of us had plenty of room for sprawling about. We spent a good half of our time in there. It was a movable palace.

Movable: for we traveled tirelessly. We crossed Siberia at a gallop pursued by packs of wolves or an America teeming with Indians; we brought succor to the Crown Prince across the bedlam of a lost battlefield. Within this imaginary framework— always determined by Bertrand—the scenario was changeless. We would arrive with grim expressions, and Pierrot, eternally the manservant with the heart of gold, would load bread, goose-liver sandwiches, and cider filched from his father's farm into the trunk at the rear. Once inside the vehicle, under the muted light of the lamps, Bertrand would address each of the two girls in turn.

"Zazanne, this will be a dangerous mission. We may never return alive. Are you brave? Are you faithful? Are you determined?"

"Yes," Zazanne would breathe in reply, in the throes of emotion, overcome with gratitude for the noble lord who deigned to cast his eyes in her direction.

"Well, then, prove it. Show!"

That was the key word. An imperative verb without a direct object. Nothing had ever been specified. The first time Bertrand issued the order, neither girl hesitated. What else would they have had to "show"?

Zazanne dropped her pants to her ankles, raised her checkered dress high above her head, and stood upright among us, her columnar thighs pressed so tightly together that her stocky nakedness lost none of its essential mystery. And none of us, quite obviously, would have wanted to know more. She was plump, with little, pear-shaped breasts, hips strong for her age, and a woman's pubis nicely furred with brown curly hair. The first time, Pierrot had laughed, nudging Zigomar.

"So that's what all the fuss is about!"

Bertrand transfixed him with a look. "What's so funny? You're nothing but a hick. All right, that's enough, you are beautiful, Zazanne."

The examination had lasted only a few seconds, and Zazanne, tears in her eyes, probably of happiness, or so I feel, nimbly raised her pants and let her dress fall. As for me, I didn't think she was so beautiful. I was somewhat ashamed at taking such pleasure in looking at her. I was waiting for Maïté's turn.

"Maïté," Bertrand said in exactly the same tones, "this journey will be perilous. Are you brave? Are you faithful? Are you determined?"

Maïté shrugged with a small rebellious smile.

"Yes," she threw at us, the way one throws a dog a bone.

"You are speaking to *me*," Bertrand's imperious voice cut in. "What is your answer?"

"My answer is yes!"

If I were not afraid that the comparison might seem ridiculous or offensive, I would say that the only other time I have heard the word "yes" pronounced so fearlessly, with such religious exaltation, was when the daughter of a friend of mine entered holy orders; she was a truly ravishing girl, and I had been invited to provide moral support for the parents . . .

"Well, then, prove it," Bertrand ordered. "Show!"

This time no one laughed. We scarcely dared look, as though so much whiteness under the feeble flickering glow of the candles dazzled us. We sat frozen to our seats as the pants fell and Maïté's dress was raised. I have tried to recreate the feelings we experienced at that moment. *My* feelings, at least, and I'm certain that those of Bertrand and the other adventurers in the omnibus didn't stray very far from my own. And it is important — now that I have found this story buried alive in the depths of my memory — that I do not confuse the child who lived it with the man who tells it, that I do not lend to the first the basic reactions of the second. We are not dealing with men. At that age, in those days, and remembering the relative naïveté of children back then, I do not believe, to put it plainly, that we had erections in the gloom of the Majorel omnibus. Under other circumstances, yes, but not there, not while looking at Maïté. We were physically capable of it, but emotion made it impossible. An intense emotion, liturgical, something carnally divine, whose roots were somewhere removed from the normal seat of such agitation and which left us pensive long after Bertrand said, "All right, that's enough, Maïté. You are beautiful."

And our heads remained in the clouds until Bertrand stuck his head through the carriage door and sang out, "Off you go, Kolb! At the gallop! Use your whip, coachman!"

Bertrand waved his arm as though ordering a charge. His bracelet glimmered in the half-light, reflecting the twin candle flames. Six feet above our heads, at the unlit top of the landau where the coachman's seat was perched, the whip cracked. We heard it whistle over the phantom harnesses into the gloomy recesses of the building.

Coming down to us from above, a voice cried, "Tchah, gee-up, Diamond! Gee-up, Topaz! Get along, lazybones, Ruby, get along!"

It was Kolb. He was an eighty-year-old child.

Kolb had been the Majorels' coachman. He had not worked for twenty years. A little old man, gnarled and dry, with a shaggy mop of white hair and still-black brows, which he constantly furrowed (it was virtually his only means of expression, save when he was addressing his horses). An Alsatian. It was said that at the age of ten, in 1871, he had fought with the partisans of the Vosges mountains against the Prussians and that he had ended up in Touraine, an orphan, after a long and lonely march. Bertrand had managed to extract a few words out of him on the subject, and ever since, the child and the old coachman had shared a complicity they felt no need to articulate. When the Majorels had locked up their coach and sold their last horses, Kolb had stayed on at La Jouvenière. Where else could he have gone? He was no longer paid—a small sum at Christmas, another on Easter Sunday, the occasional tip—but he was cared for when he fell ill, he was fed in the kitchen whenever the sole remaining maid was so inclined, and he slept in the stables among the memories of his horses, helping around the grounds, cutting firewood and doing what he could to arrest the decay of the park and driveways. He undoubtedly possessed a first name, but everyone at La Jouvenière simply called him Kolb and spoke to him as if to a child. He was childlike rather than senile. He had probably always been so. It was he who opened the carriage door to us, in his grime-stiffened cap, before scampering like a monkey to his lofty perch.

"Kolb!" Bertrand would shout up at him through the lowered window of the door. "Can you hear the wolves?"

Variant on the scenario: "They're shooting at us, Kolb! Did you see where the outlaws are hiding?" Or again: "They're after us, Kolb! How many riders? Can you see them?" Then the whip would crack more merrily than ever, and Kolb, upright on the running board of his seat, coaxing the horses with his voice, would leap about until the ancient vehicle shuddered from his exertions. It was a *Fantastic Cavalcade* before its time. Later on, watching that movie a first and second and then a third time, I couldn't help remembering. Inside the bus everyone was at action stations. We pointed our rifles, our child's rifles, we aimed them and shouted, "Bang! Bang!" and pretended to reload feverishly.

"My friends!" Bertrand shouted. "Let us sell our lives dearly and protect the honor of the women!"

I don't for a moment believe that "the honor of the women" had any specific meaning to us, other than that of refusing the imaginary highwaymen the sacred vision of a white belly reserved for us alone (which is actually not bad, now that I think of it, and which made for healthy emotions). At least we knew why we were fighting.

"I'm frightened! Protect me!" moaned the hapless Zazanne, huddled in a corner of the bus.

Then she threw herself into my arms, or Pierrot's, or Zigomar's—it was her role, the only one she could play with dignity—and more rarely into Bertrand's; he was too absorbed by the battle. He maintained a steady rate of fire (bang! bang! bang!), while Maïté, crouched at his feet, reloaded her lord's rifles and passed them to him with a nobility of gesture worthy of an allegory of war. It was the destiny they had elected to share. Maïté didn't play faint swooning women. She would never have agreed to, and Bertrand would never have required it of her. Between volleys, the rest of us would console Zazanne, who would purr like a cat. Caress her hair, join hands moist with emotion, a brief and clumsy embrace, then back to the fight. But Zazanne was not to be forgotten. Particularly not by me. If I spent too long at the carriage door, my rifle in my hand, playing the heroic musketeer, Bertrand would shout at me, livid, "How is Zazanne?"

Instantly Zazanne would chime in, "I'm frightened! Protect me!"

And I would climb down to the seat to console Zazanne, shamefaced, stung, submissive. It was my part, the one that Bertrand, deciding that I wasn't much good for anything else, had assigned me once and for all. It is true that I was a mediocre shot, that I feared firearms, the bottom of the garden at night, the river when I was out of my depth (even though I could swim), crossing darkened hallways where they insisted on dragging me to make me ashamed. Poor, tender, pitiful age . . . It is true that I didn't know how to fight, that I didn't like to fight, to roll about on the ground in the dust in defense of some obscure point of honor, that a visceral fear lived inside me, and that on several occasions I had shown myself incapable of hiding it. I would make great efforts to redeem myself in Bertrand's eyes, or Maïté's. Every time I managed it, through some miracle of willpower, I would fall flat again at the very next hurdle. That was the cross I bore, but it is also the reason Bertrand showed such friendship for me, more than for Pierrot and Zigomar, and why I clung to him. At least I existed when I was near him. So when he packed me off to console Zazanne I went! None too sure how to go about it, in the end I kissed her neck, while my hand, quite by chance, found her soft little breasts, and my heart stopped beating.

"Master Bertrand! Master Bertrand!"

It was Kolb. He called us all "Master" or "Miss," a vassal to his suzerain, not a servant to his master, or at least that is how we understood it.

"Master Bertrand! They're running away!"

"Cease fire!" Bertrand cried.

And Kolb calmed his horses. "Easy, Diamond. Easy, Topaz. Easy, Ruby, you lovely girl."

The bus finally at a halt in some deep forest clearing, we tallied up our exploits.

"I got five!" Pierrot announced.

"Three for me!" Zigomar chimed in, his shortsightedness making him modest.

"I think I missed one," I added pathetically.

Clumsy, pacifistic, cowardly, I remained true to my assigned role, and they laughed kindhearted laughs.

"And I killed their leader!" said Bertrand, looking at Maïté as though he were throwing the victory spoils at her feet.

Bertrand always killed the enemy chief. Not one of us would have risked stealing that signal honor from him. Our lovely little children's society worked smoothly . . . Sometimes he would be wounded. A bullet in the arm, an arrow in the thigh. Maïté would unpack her Red Cross nurse's kit and make him a nice bandage. After which we would refresh ourselves, sandwiches, bread, cloudy cider quaffed from battered silver goblets; it always went to our heads a little.

"Kolb, old soldier, are you hungry?" Bertrand would ask.

Kolb was always hungry, and thirsty. We would pass him enormous pieces of bread, a whole bottle, which he would guzzle down without a glass, on his seat, with us in spirit but isolated, true to his own ideas of decorum and of his rightful place. We couldn't have played without him. Old child that he was, his *adult* presence lent authenticity to our games. All that could last a whole afternoon, but never did Aunt Sophie Majorel say a word about it. She would, however, express surprise.

"But what do you get up to in there?"

"We play, Aunt!" Bertrand replied, smiling with all his teeth.

Zazanne fidgeted a bit and blushed, but Aunt Sophie pretended not to notice. What Bertrand did was fine. Bertrand could do no wrong. In which she was correct.

As for my aunt Melly, when I returned out of breath from pedaling along the road and only just in time — but in time nonetheless — for dinner, she wouldn't question me either, for I always gave her the same reply, "I was with Bertrand!"

Just as I would have said, "I was with the king!"

That was enough for her.

I have checked the date: the Soviet-German Pact was signed August 23, 1939. It was therefore on that date, in late afternoon, at the end of the first (and last) big hunt of the summer, that we met at La Jouvenière, the six of us sitting grimly on the omnibus benches, hands folded in our laps. Kolb had been asked to unhitch the horses and to sit peacefully on his perch. We were not embarking—not yet—on a warlike mission. There would therefore be no opening ceremony, and the girls, the hems of their dresses properly pulled down past their knees, silently waited for the boys to speak. The silence lasted. If I remember correctly, I believe that without actually abandoning our game playing, we were conscious that the game had undergone a fundamental change. From now on our imaginations would be feeding on reality.

"Right! Report!" said Bertrand.

That was another of his expressions, no doubt stolen from his father's military vocabulary. All that was missing was the trumpet sounding inside the lonely fortress.

"What have you heard?" he asked us.

"My father told Uncle Léonce that this time the balloon really has gone up," said Zigomar, "and that it'll be here within the week."

"What will be here?" asked Zazanne.

"War."

"Oh!" she cried, hiding her face in her hands. "I'm frightened! Protect me!"

"Shut up, idiot!" Bertrand snapped. "And you, Pierrot?"

"Same thing. Mr. Raymond told Mr. Armand that mobilization can't be far off, because the Germans will soon be invading Poland."

"Are the Germans strong?"

"Nobody knows."

"What about us?"

"If you had seen their faces! Especially the young ones, the ones that have to go first. I don't know if we're strong, but I know we're unhappy. They don't think it's funny at all."

"And the Germans? They think it's funny?"

"Apparently. They sing and they goose-step through the streets making a lot of noise with their boots."

"The swine!" said Bertrand, his eye gleaming with pleasure, glad to have thus defined the enemy that he, Bertrand Carré, well-brought-up French youth, had elected to face . . .

"And you?" he asked me.

"I've heard my cousins talking. Three of them are officers. It's true, they don't like war."

"There you're a good judge of character," said Bertrand. "Well then, we are going to do it ourselves!"

"Good idea!" I said, trying to enter into the spirit of the game. "Here's to war!"

The mere courage of words. It was bad. It was pitiful. Coming from me, it rang horribly false.

Bertrand merely went on, "We will begin by getting ready for it. Tomorrow, you will all repaint your bicycles khaki. How much money do we have?"

Everyone emptied his pockets. With the wages of the four beaters, it added up to a war fund the likes of which we had never seen before.

"We'll go to Magloire's tomorrow," Bertrand went on. (Magloire's was Petit-Bossay's grocery–snack-bar–general-store–tobacconist–gas-station.) "We'll clean out his entire stock of Bosquette 5.5's." (The brand name and caliber of our rifle cartridges.) "And we'll begin training on Blue Island. If we have any money left, we'll buy provisions, canned goods, biscuits. We must be ready when the time comes . . ."

He already had a plan.

"IT'S NOT A stream — it's a river! Didn't you know? The Mulsanne flows into the Creuse, which flows into the Vienne, which flows into the Loire, which flows right out to sea. The Vikings launched their attack from the Mulsanne when they wiped out Beausoleil and beat the local pheasant hunters to their knees. If we wanted to, we could go straight from Blue Island to America!"

Such were Bertrand's words when, soon after his arrival among us, we took him on a picnic to Blue Island for the first time. He had immediately pushed back our boundaries, shattering our habits, lending an unexpected freshness to the humdrum workings of our imagination. Blue Island was a delectable little spot, covered with clumps of trees and tufted with wild-rose bushes and a small stand of poplar, with fallen trunks we would straddle in order to fish, and with tiny beaches of white sand. A hundred yards in length, more or less, by twenty or so yards in breadth, lying flush with the water between the unequal arms of the Mulsanne, the northern arm relatively deep and wide, the southern arm narrow and almost overgrown, for the trees on its banks met and intertwined their branches above it. A brook swollen to small-river size a few miles upstream, the Mulsanne, like many of the region's waterways, resembled a miniature Loire, the new direction of each new meander offering a new self-contained universe, harmonious, blessed with a divine-seeming calm, a paradise of greenery, sand, and water. Ours was bounded fifty yards to the east of Blue Island by a half-submerged retaining dam and the crumbling fragments of a wall around an abandoned mill, and a hundred yards downstream to the west by another, smaller island called Saint-Cyrin, where the remains of an old chapel stood amid the underbrush. Before the irruption of Bertrand Carré we had asked no more of Blue Island—and I don't know why we called it that—than fun and games without theme or plot: fishing for minnows, hide-and-seek in the undergrowth, bathing at the westernmost point of the island where the river's two arms joined to form a nice stretch of smooth water, knife throwing in the sand, and hunting for the small birds at which children in those days, unexposed to ecological considerations (the word itself had only just been invented), would fire without compunction. No grounds for comparison with Robinson Crusoe, or with Pierre Savorgnan de Brazza, the African explorer, or with Fenimore Cooper's deerslayer. With little picnic baskets fixed to our handlebars we would arrive in orderly fashion along the trail, a fairly wide gravel road that was almost never used—it led, to no purpose, in the same direction as "the main road" a mile to the east—and that traversed the island across a pair of iron bridges.

I see it quite clearly now, our island, alive once more under the luminous summer sky of Touraine. I have described it in such detail because it was here, on the first bridge, that in June of the following year . . . But let me turn now to those two bridges with their iron roadbeds, iron arches, and massive iron bolts. In the twenties a stretch of line from a private railway had crossed the bridge, heading from Châtillon-sur-Indre toward a quarry south of the Mulsanne. The weight of the loads it carried explained the solidity of the bridges and the width of the trail that had subsequently replaced the tracks: for the quarry had been shut down; the Depression had swept everything away; even the rails and railroad ties had been ripped up and sold.

And finally, to complete our children's world, the ground for four or five hundred yards north of Blue Island was covered with thick undergrowth; at its extremity, thrusting upward from a low hill overlooking a bend in the trail, was the stump of an ancient crumbling castle keep. We were forbidden to climb there for fear of "breaking our necks," a decree the aunts lifted as soon as Bertrand arrived on the scene and wrapped them around his finger. It was known as the de Réfort tower, and it belonged to Maïté's parents, as did Blue Island, the mill on the Mulsanne, Saint-Cyrin chapel, and all of the neighboring territory. But we called it simply "the fort," the way we said "the mill" and "the chapel." From its truncated heights we could see the gray ribbon of the trail vanishing among the trees into the woods to the north and across the two iron bridges to the south.

That was Blue Island. And Bertrand preempted it the second he set foot there. Maïté had ceded him that honor as if bringing it to him as a dowry. Love at first sight! He had fallen in love with the island at first sight. The first thing he had done was pace it from end to end, an administrative staff of admiring kids on his heels, all rediscovering the island through his eyes. He assessed the height of the trees, climbed to the very top of a couple of them, moving steadily from branch to branch, picked out choice spots for tree houses, then leaped nimbly to the ground before our astonished eyes. When we reached the sandy western tip of the island he stopped at the very edge of the woods, arms spread to prevent us going farther.

"Stay right here!" he ordered.

The river seemed to have arrayed itself in finery that today, as I question my memories, I would be tempted to call feminine: a miraculously soft light, the whispering of the water, the reflections of sun and shadow, the dazzling white of sandbars surfacing here and there like motionless swimmers, the almost hypnotic gleam of the moving waters, looking like hundreds of intensely

green and translucent eyes, and the song of a single nightingale
divinely trilling into the silence – for all the other birds had
mysteriously fallen silent. So had we. Maïté, Zazanne, Pierrot,
Zigomar, and I, we waited, arms dangling, motionless, mute.
Bertrand had removed his shoes, his shirt, and his trousers and
stood before us in swimming trunks. He walked soundlessly
toward the river, leaving the imprints of his bare feet in the sand,
little by little marking out the distance separating us from him.
On that day it was enormous. I spoke at the beginning of this
story about Bertrand's beauty, the princely elegance of the lines
of his body, the long, nearly imperceptible muscles – in his
shoulders and back for example – graceful rather than powerful,
even though he was already robust. Obviously we could not have
articulated anything along these lines, but I believe that it was a
kind of confused adoration that we felt at that moment. Then he
went on into the current, wading another ten yards or so along
the submerged sandbank until the water was at mid-thigh, and
there he stopped, standing, his back to us, planted squarely on
his feet, facing the river as it glided by like a sinuous green
animal. He was now quite far from us, and we could not
distinguish the details of his movements with precision. His
trunks sat low on his buttocks, his arms were joined in front of
him, and the tilt of his head indicated a function we all knew
well.

"He's pissing!" Pierrot exclaimed.

From the solemnity of the ceremony accompanying such a
commonplace act we inferred that he was marking his territory,
as do dogs. He was most definitely marking it, but not in the way
we believed. That, I understood later. After which he gave a great
whoop, a phenomenal howl of joy, plunged into the flowing
stream, and swam furiously, kicking up showers of water that
sparkled in the sun. Finally he turned back to our motionless
little group.

"So, are you coming in? Yes or no?"

It was all we needed. We were in bathing gear, the girls in
chaste suits hardly cut away from their throats, a bit more at the

arms and not at all in the thighs, about as transparent as lead when wet, Maïté in sleek black, Zazanne in gaudy scalloped red.

"You don't have a black suit? Something plain?" Bertrand had asked.

"Why? You don't like this one?" the poor girl had responded, crestfallen.

"Black alone suits the skin of a girl in a bathing costume. Get that into your head once and for all!"

A remark that would have entertained the aunts and that left us speechless. The very next day Zazanne complied, and it is true that black transformed her. Despite her heavy thighs, her big bottom, her thick lips, it made her almost elegant.

Afterward, we amused ourselves like madmen — and not in the hackneyed sense of the word. We were not in our normal state. Innocence and ingenuousness were coupled with a new feeling, an exaltation both physical and spiritual. Our souls were laid bare. I have some difficulty flushing from the shadows of time and transposition the childish reality, for we were not truly conscious of it. First we swam to Saint-Cyrin Island in a group, arms and legs instinctively harmonizing their movements with the stream, like some carnal tribe migrating with the changing of the moon, and Bertrand took possession of the island, Blue Island's colony and outpost. Standing in water halfway up his legs before gaining the bank, he asked us to wait behind him.

Pierrot breathed to me, in a low voice, "He's going to piss again!"

Which he seemed to be doing, his stance once again indicating that he was performing that familiar act. Maïté let her gaze stray, a stranger, for the moment absent. As for Zazanne, forehead boiling, scarlet, I believe that she had more or less discovered, before us and by instinct, what it was that animated the proud young man. She wore a foolish and secretive smile worth a thousand words . . .

Then we ran to the ruined chapel along a footpath overgrown with brambles; they streaked our naked calves with blood, the leafy branches splashing our bodies with pearly drops of dew, the girls' hair like never-ending fountains flowing the length of their backs.

All that remained of the chapel was an arrangement of stones on the ground forming the outline of a Romanesque apse. In the time of the barbarian invasions, the dark ages of Rome's imperial decline, a hermit had lived here, part lost soldier, part visionary bewildered by Roman persecution of Arianism, forgotten for years, and finally struck forever from the pantheon of those canonized by popular fervor.

"And now," Bertrand said, "let us pray!"

Zigomar and I were ready for anything. We would with confidence have recited the rosary on our knees. Every scout sees the hand of God in nature, and Zigomar and I were good little scouts in 1939, members of troops recruited from well-to-do Parisian families. Pierrot scratched his head, prey to the fundamental doubt of the peasant.

"Not the rosary, fools!" Bertrand said.

Joining his hands, he improvised, "Saint Cyrin, gird our loins, give us courage, rage, and carnage, and give the damsels respect and undying homage. Amen."

A spontaneous nursery rhyme. Gifted and inventive, Bertrand had been influenced, I believe, by the arrangement of sounds and the music of the words rather than by their actual meaning. For "damsel," for example, I was forced to sneak, that very evening, to Aunt Melly's dictionary (1. *Archaic:* A young woman or girl. 2. *Fam.:* Virgin woman), which didn't help me much. Neither was Bertrand showing disrespect for religion, merely a disrespectful attitude. His own way of praying. For he had sincerely meant his prayer. Today I no longer doubt it. No matter that we had remained silent, incapable of imitating him, memories of catechism and school prayer obstructing our throats. No one had ever taught us anything like that. Maïté alone joined in, her sugared tones already betraying the vowels of the seventh arrondissement. Standing on the tips of her bare toes, legs together, she raised her arms skyward in a gesture of invocation, slimmer and more delicate than ever, beautiful, the damsel incarnate . . . There was less affectation there than we thought. A moment's inspired grace. I had never seen a girl pray in such a manner, barely out of the water, skin glistening and wet. Or in such words. She omitted none of them, and they were charged with mystery by the mere movement of her lips. After which we bombarded the improbable Saint Cyrin, patron of the Six-Four-Two (six kids, four boys, two girls: our gang), with stones, and turned to other games.

We built rafts, and tree houses at altitudes that seemed dizzying to us. We blazed narrow trails—our "lines of communication"—with hatchets through the thickets and wild-rose bushes. Wearing American Confederate caps with cardboard peaks bought at that summer's fair, we played at capturing or defending the iron bridge, and, the battle lost or won—depending on Bertrand's mood—we paid military honors to our fallen. To our one fallen. Who was always Bertrand. He had an eye for symbols. That role, too, he reserved for himself. He would lie beside the path in the recumbent posture of the dead with a flag—sketchily put together by the girls—covering his body, red with the blue and white Saint Andrew's cross of the Confederates, his gray Southern cavalryman's cap on his chest, hands clasped, eyes closed, as though he had just given up his life in a successful defense of a bridge over the Potomac. It was rather impressive, that old-fashioned picture of heroism. Zigomar, Pierrot, and I would present arms, while the girls were asked to show their grief *with dignity*. For Maïté, that went without saying. Hers was an aloof sorrow. Except for the day when (for the sake of variety, I believe) she stepped up to the corpse, one hand lowering her pants and the other raising her dress, giving us all the briefest flash of whiteness, nimbly, gallantly, in broad daylight, saying, "Since you are dead, you will never again see *this!* It was made for you." Zazanne had no genius for improvisation, but she wept, almost sincerely and alas without dignity. Then the fallen hero winked at me, telling me in a mocking tone, "Console her!" That was how it always ended. Then we jumped on our bicycles to end the afternoon at La Jouvenière, in the old dream bus, to wage war on the brigands along the Great Wall of China.

"Gee-up, Diamond! Gee-up, Topaz! Come on now, lazy-bones, Ruby, come on!" Kolb would cry.

Ah, the real and beautiful lives of children . . .

To start with, we got ourselves soundly shouted at for our clumsily painted khaki bicycles. Even Aunt Melly was visibly annoyed. "Whose idea was this? Bertrand's, obviously! Do you really think that now is a good time for such tomfoolery?"

But that was nothing compared to the anger of the young uncles and the older cousins, themselves doomed to khaki the instant the ordeal began. The rooster-feathered rabbit hunter, my cousin Raymond Bonnadieu, was having a particularly hard time digesting the situation. Determined to take it out on us, whom he encountered as we pedaled along a Beausoleil trail, he erupted from his Renault Celtaquatre, fuming with indignation.

"Another of this little prick's ideas! His father a squadron leader" — he was naturally talking about Bertrand — "and he's playing the fool at a time like this!"

He repeated the phrase "at a time like this!" several times, and at a time like that, it was true, things weren't going well. They had begun by calling up officers on leave. He himself had just received the order by telegram. He followed by attacking me.

"And you! Chicken liver that you are! Yellow belly! Gutless! You think you look tough? Ah! It would do you some good to go to war . . ."

"And does war suit you so well?"

It was Bertrand, thoughtfully leaning on his handlebars, his smile insolent. My cousin must have come close to exploding with rage.

"You little . . . You little . . . There'll be men killed, wounded, crippled for life, families destroyed, widows, orphans . . ."

For my part, I thought he was pathetic. He most assuredly was, but Bertrand went on smiling as my cousin continued, "Villages in ruins, misery for everyone! Do you think this is a child's game? Do you think I'm playing a game?"

"If only you were," Bertrand sighed.

It was Cousin Bonnadieu's stupefaction that saved Bertrand from a well-deserved set of boxed ears. He got back into his Celtaquatre.

"And don't let me run into you again! The farther you stay out of sight the better for everyone . . ."

By next day he had left Beausoleil, in khaki, tears in his eyes — he was a newlywed — and two silver stripes on his sleeve. I have already told how Lieutenant Bonnadieu was so bravely killed at Arras; I do not feel entirely at ease describing the scene above. Bertrand, however, showed no remorse when the news reached the family, in June of 1940, a few days before the appearance of the Germans in Touraine. It fit his haughty logic.

Stay out of sight. That is exactly what the Six-Four-Two did for the next ten days, from August 23, date of the signature of the Soviet-German Pact, to September 3, 1939, the day war was declared. They were good days. We roamed freely between Blue Island and La Jouvenière (the omnibus), leaving early in the morning on our khaki bicycles and returning late for dinner, escaping the shroud of sadness and apprehension that had fallen over Beausoleil, over La Celle, La Guichardière, La Cornetterie, the homes of the Bonnadieus, the Réforts, the Majorels, the Durands, the Lavallées. They paid one another frequent and sorrowful visits, made interminable phone calls, and sat glued to the radio in order not to miss the news. They were not acting in a drama, they were enduring it, like all of France for that matter, if my memory holds true; hence the heavy, resigned atmosphere, the funereal meals around tables where the number of guests shrank day by day as uncles, cousins, and friends left to rejoin their regiments. Even the aunts had renounced their usual practice of planting sarcastic barbs in the infinitely weary humps of their menfolk. And when my poor uncle Sébastien Lavallée — husband to Aunt Melly, who had poured such scorn on him the day he wore his "Boer costume" — had in his turn been obliged, with a thousand comic contortions, to squeeze into his old, faded reserve captain's uniform (which at once split at the armholes) in order to go to war fifteen miles to the north, in

Loches, taking command of the village square (Loches square! fortified in the Middle Ages! from which, in any case, he returned to spend every Sunday with his family, his feet warm and dry), this time my sharp-tongued aunt did not permit herself a smile, none of her customary barbs, not the slightest look of amusement. She quite simply sobbed. Her hero inspired her to tears. Not the shadow of a sign of exhilaration, of pride.

Oh, the doleful departures I witnessed on the front stairs of so many of our neighboring châteaux! They lacked only embroidered funerary crepe bearing the deceased's initials in gold letters, M for Majorel, R for Réfort, B for Bonnadieu. No longer dashing, the little lieutenants left us; so did the old captains, already exhausted by four years in the trenches twenty years earlier. They had no idea what was waiting for them this time, nor why they were fighting; they went from force of habit and a sense of duty. The barons and vassals of the French republic. Reflections of the French body politic . . . They embraced their offspring and their wives, already bowed down by sorrows, already widowed; they shook hands with their gamekeepers, who twisted their mustaches with emotion, and they left, dazed and resigned, for their improbable posts on a surreal battleground. We, the children, opened incredulous eyes. What! No one was singing! No one bellowed a patriotic "To Berlin!" the way they had in 1914 at the Gare de l'Est in Paris; we had all read about it in our history books. At the Châtillon-sur-Indre train station, where the aunts' convertible C4's dumped our menfolk like whimpering orphans, there were no buoyant slogans painted on the railway cars. Just a desultory and mournful consumption of sandwiches and white wine among the common soldiers and a nearly sepulchral silence in the officers' compartments.

The departure of officer-cadet Jean-Louis de Réfort, Maïté's older brother, was a model of Shakespearean mystification. I think he was truly keen to go. But his parents — they felt dishonored! Aunt Octavie, looking twenty years older, mumbled. Uncle Gaetan de Réfort, judged unfit for service but certainly not senile at forty-five, foolishly saluted his proud son, magnificently well proportioned in a shining new uniform, and spluttered through wet kisses, "Come back to us quickly, my boy. Come back to us . . ."

And Bertrand, hearing their words, mustered us with a single glance, "Do people come back from war? Do they?"

"But he's my brother, after all!" Maïté protested.

"Exactly! Brothers can have widows too."

And on that enigmatic note, he took us to practice our marksmanship on Blue Island.

Again, let us be fair. Officer-cadet Jean-Louis de Réfort left his bones on the bridges of Saumur, commanding a company of cadets.* Captain Pierre Durand, Zigomar's young uncle (we saw him leave as well, tearing himself from the moaning embrace of all the Durands of La Cornetterie), gave his life in a vineyard near Dijon beside a lone 37mm field-gun manned by Senegalese. It was not the aristocracy that lost the war of 1940. Nor was it the true French middle classes. Nor was it truly the people of France. In truth, it was very few people. Or rather it was a delinquent soulless mob, all those who sang Momo's songs and believed they could drown Hitler's High Masses at Nuremberg with their crowing. I can say that now, as so many others have. But at that time we children knew nothing of such things.

We had hauled an old iron trunk up into one of our tree huts, the one in the big beech near the northern bridge to the island. It contained our "strategic reserve," a term you heard on the radio a lot back then. It was where we stored our Bosquette 5.5's, wrapped in tarred paper, as well as sugar, chocolate, cookies, and three cans of sardines. We had quickly depleted our war treasury and were forced to fall back on pilfering whenever the cooks' backs were turned. Soon we broadened the scope of our larceny until we had a bit of everything in that trunk: balls of string, tincture of iodine, matches, Michelin maps, canned peas, bandages, a compass, a gas-burning camp stove.

---

*Saumur was an authentic gallant last stand. — Trans.

When the ever-practical Pierrot asked what we were going to do with it all, Bertrand replied, "They're supplies for the flying column. We may need to travel far . . ."

An interesting coinage, doubtless dug up from the colonial vocabulary of Squadron Leader Carré . . . While awaiting the departure of the "flying column," we spent the better part of our time exploring the contents of the "strategic reserve." We ceaselessly checked and rechecked the inventory, everything, thanks to me, duly recorded in a schoolboy's notebook, with oohs and ahs of pleasure whenever one or the other of us, with a conspiratorial expression on our face, threw some new object into the communal pot, an individual mess kit or an old pair of hunting glasses forgotten in some attic. A whiff of faraway conquests escaped the open trunk. Obviously our models, our real-world sources of inspiration, were out of date. Children worthy of the name are always out of date. Then we would clamber down from our tree and spread out all over the island, exuberant, excited, unstable. Stampedes to the ends of the footpaths, frenzied swims in the Mulsanne, volleys of shots at old pieces of crockery from our Bosquette 5.5's. All that expenditure of physical energy to calm our impatience. Zazanne and Maïté had to drop their pants more than once to reestablish a natural order in the turbulent pattern of our thoughts. Those little triangles of tender flesh, covered up as quickly as they were revealed, represented the only cosmic truth that held firm amid the collapse of everything else. At least that is how I explain it to myself today. From the deathly silences around the radios in the evening at the news hour, from the increasingly somber conversations that followed, we understood that the world was crumbling, and I remember very well what we thought of it all: if it had been in our power, as children, to do it, we would willingly have given the tottering edifice a little shove! Bertrand summarized it all for us one morning.

"When are the fireworks going to go off?"

The next day, September 1, they went off. At four-forty-five a.m. Hitler invaded Poland. The Reich's armored divisions — people didn't say "Nazis" then, it was not part of the age's vocabulary—drove the hapless Polish cavalry, caught by surprise in the throes of mobilization, off the field on the very first day. After its ultimatum was rejected, Great Britain declared war on Germany at eleven a.m. on September 3. The French parliament wrangled over details in a long emergency session that took another six hours, so France did not follow suit until five p.m.

The clanging of the village bell, carried by the wind from the church tower at Petit-Bossay, surprised us as we pedaled along the trail to La Jouvenière. We leaped from our bicycles. In those days church bells were not worked by electricity, and the chimes still had a soul. Great ropes dangled from the bells, passing through holes in the wooden floor of the tower to be rung by invisible hands far below. The joy (chimes), the mourning (knell), and the fear (alarm) that the bronze throats expressed were in tune with the heart and soul of the bellringer. At Petit-Bossay, Magloire the grocer, who also held the office of sexton-sacristan, rang and rang as though his life depended on it. The staccato clanging of the village's only bell, rapid, breathless, was not intended to evoke courage or patriotic pride. It was a cry for help, and its rhythm was desperate. Our hearts felt tight in our chests. That ringing reeked of fear. We looked sheepishly at one another, not knowing what attitude to assume. Zigomar mechanically removed his cap and made a hasty sign of the cross, quickly followed by Pierrot and me. A prayer? But for what? Victory? That was a word we had not once heard mentioned in the pathetic talk of our elders. The uncles were going away to war, but not to win it. Then why? We had no idea . . .

Bertrand pulled us out of that terrible state. With his Confederate cap in his hand, heels together, nose pointing to the blue sky where the harsh clanging of the alarm bell seemed to reverberate, he shouted out spiritedly, "Saint Cyrin, gird our loins, give us courage, rage, and carnage, and give the damsels respect and undying homage." Then, mounting his bicycle, he cried, "Yee-hoo! To the omnibus!"

We pedaled like maniacs, the alarm following us all the way. Kolb was waiting for us in the shed, eyes shining, with a most unexpected antique blunderbuss slung across his back, probably the very one that had fired on the Prussians in the Vosges in 1871, when he was with the partisans. He had already lit the bus's dim lights.

"It's war, Master Bertrand! Where are we going? I've hitched 'em up."

"To Poland!" Bertrand announced with the ardor of a Dumas hero, adding, "To Nesle Tower!"

For three days the radio had been pounding our ears with the exploits of the ill-fated Polish army, succumbing gallantly under a hail of shot and shell. All the talk was of cavalry regiments madly charging German tanks at full gallop, lances with red and white pennants at the ready as though for a tournament, borne along on trumpet blasts that faded only when the very last of these centaurs was brought down on the immense flat plain that swallowed the blood of the corpses littering it. Something like that, anyway, and there was without doubt some truth in it. The father of one of my friends, a colonel of Polish lancers, was cut down in exactly that fashion, at the head of his regiment, his commander's riding crop in his hand. And all this time the French troops, active regiments raring to go, champed the bit like extras at the opera. The uncles and cousins methodically stuffed their packs with laundry and other unwarlike parapher-nalia. They did not seem in the least impatient. We wanted to shout at them, "Come on, hurry! It's going on without you . . ." As for us, we were there already. Our hearts were beating for Poland, where brave men were dying, while here . . . But no one saw these storms swirling behind our impassive children's masks, no one heard our secret judgment of our elders.

Inside the bus, its doors closed, we lived the moment in-tensely. The theme of that day's game: to force a passage through the German horde and die in the ranks of the last gallant Polish regiment. Bertrand announced gravely, "Zazanne, this mission is doomed. We will not return alive. Are you brave? Are you loyal? Are you determined?"

I felt disturbed. I went red. By my side, Zigomar swallowed painfully. Something in Bertrand's tone of voice told us that if he

was still playing, he was playing a different game. Maïté was
watching him closely.

"Shall we do without the girls, just this time?" I ventured.

The answer was sharp as a knife.

"You idiot! I told you, you're going to die! What's supposed to
comfort you as you breathe your last? What will you take with
you from your unimportant little life? Who will your last thought
be for?"

"For my mother?" I said, timidly.

This seemed to me perfectly obvious. I had read that on the
battlefield, in 1914, the dying invariably called out "Mama!" as
they breathed their last.

Bertrand crushed me with his disdain.

"You idiot," he repeated. "Your mother . . . My mother . . .
Mama . . . We're not kids anymore, you fool!"

He was like an acrobat who hurls himself onto a trampoline
and bounces back up through a paper hoop on the other side,
transformed. Child on one side, man on the other. Maïté went
on watching him, her gaze deep and unwavering. In a flash of
white cotton, her pants were the first to fall.

"So? Idiot! Now do you understand?"

What was there for me to understand? I was adrift, abandoned,
out of my element. The memory brings back other dismal
fiascos I endured later on, in the company of willing girls who
had judged me wrongly. Maïté's inaccessible beauty, I remember
that as well . . . At the beginning of this account I said that I had
erased Bertrand from my memory. This scene was a contributing
factor.

Vaguely aware that there was something beyond her grasp at
stake, Zazanne waited.

"You are beautiful, Zazanne," said Bertrand. "That's enough,
now . . ."

The poor girl began to sob. She too, perhaps, had passed
through the paper hoop, but without understanding what had
happened, a great bundle of soft, sweet flesh discovering an
emotion that eluded both heart and mind. Maïté was still looking at Bertrand. Zazanne, the others, and I no longer existed.

"You are beautiful, Maïté, that's enough."

Halfway opening the carriage door, Bertrand hailed the coach-man, Kolb, who was already cracking his whip.

"Off you go, Kolb! At the gallop! To Poland!"

I still hear those words. I repeat them to myself, and in the background I still hear the faraway alarum that reached us, muffled, through the walls of the barn, for Magloire the grocer, as if attempting to ward off fate, went on with his ringing for almost an hour. I still remember every single one of Bertrand's inflections, even after all these years. His tone was false. He was trying too hard. Our game, initially fed by this brush with reality, was now being destroyed by it. From now on all Bertrand's behavior, all the way to the end of this story, represented a sublime attempt to piece back together the fragments of the dream. Kolb's voice drowned the alarm bell.

"Tchah, gee-up, Diamond! Gee-up, Topaz! Get along, lazybones, Ruby, get along!"

The old child was still playing. In his eighty years he had heard that alarm bell three times. The three wars were jumbled together in his mind. Zazanne moaned submissively, "I'm afraid! Protect me!"

"Too late! This isn't the time!" Bertrand snapped to shut her up.

With the frenzy of the damned, we fired away from the open windows of the bus, shouting, "Bang! Bang!" Then a real shot rang out, not the ridiculous coughing noise of our Bosquette 5.5's (which we never really used except on Blue Island), but a dry violent bark followed by the strong smell of gunpowder. It was Kolb and his blunderbuss.

"I got him! Master Bertrand, I got him!"

We never knew who it was he "got," for Bertrand, with a weary gesture, put an end to the Polish expedition.

"We're not kids anymore," he said. "This is meaningless."

No matter! Symbolic, childish, clumsily carried off, it was nevertheless the only reinforcement poor Poland ever received from her Western allies. As I write this, I raise my glass. Hero-ism, in 1939, deserves a toast . . .

For a good fortnight we didn't see Bertrand again. We saw evidence of his passage on Blue Island, but he made sure we always missed him. We spotted him just once, pedaling at top speed through open country with Maïté. All around Petit-Bossay, farms and châteaux had emptied of their men. Emperor Léonce Bonnadieu, master of the hunt, supreme strategist of the partridge killers, went with his four tarnished stripes to rejoin an improbable horse-artillery regiment, which landed him, without undue violence, in a Pomeranian POW camp for officers for the next five years. The same unjust fate met Alphonse Durand, Zigomar's father, a division medical officer who, hoping to be demobilized in Limoges, ran into three German motorcyclists who packed him off to Germany, along with thousands of officers and enlisted men, to look for Pétain there. The only men left were Uncle Gaetan de Réfort, in La Guichardière, and the solemn and endearing Armand Majorel, Bertrand's uncle, in La Jouvenière, who was absolutely the only man to suffer because he was five years too old to kick the Germans out of France. All the cousins were gone. Four million mobilized: quite enough to bleed an underpopulated country white. As for Pierrot's father, he had simulated a game leg in order to pamper his fat pigs; they amply repaid his attentions during the four lean years that followed. Aimless and morose, we roamed about on our absurd bicycles.

Then Bertrand reappeared with Maïté in tow. It must have been September 20. He made the rounds of the homes, and the aunts, smiling once more, felt that the world had resumed its ticking.

He said to me, "Meeting on Blue Island."

"Where have you been?" I asked him. "Giving us the cold shoulder?"

"I've been training."

And in fact, hurtling along between the lime trees, he seemed to be pedaling with the ease of a champion.

BY THAT DATE (September 20, 1939), communiqués rarely mentioned anything but the activities of the "Free Corps." These were limited to a wooded zone straddling the Saar frontier between the Maginot and Siegfried lines. Poland no longer existed, but even so France still had a war on its hands it didn't know what to do with. To disguise this reluctance to fight and to avoid a complete loss of face, the task of defending French military honor had apparently been delegated to a handful of dogged and aggressive heroes in an area not unlike an arena, a cockpit for combat sports called the Warndt Forest, in German territory west of Forbach, where they were free to burn off their excess energy with no risk of reviving the dormant war or of endangering the nation.

The Free Corps were on everyone's lips, and with good reason: they were the only ones fighting. All volunteers, drawn from disparate regiments, they were the flower of the French army. The results of the selection process that produced them speak for

41

themselves—just a few hundred of these champions, the cream of the crop, out of a total of four million. Glorious scapegoats. They had their own heroes, their mystique, their feats of arms. They were decorated before their comrades at the front. They also had their troubadours. Everything the French and foreign press possessed in the way of war correspondents descended on the Warndt Forest the way anthropologists today swarm over the last surviving tribal groups. The best writers, the best photographers were assigned to them. And they deserved it. The magazines were filled with their exploits. In fact it was one of those magazines—*Match,* I believe—that Bertrand now pulled from his pannier along with an assortment of Michelin maps.

"Flying column leaves first thing tomorrow. We're going to join the Free Corps."

That news caught us flatfooted. Although school had been postponed for a week while retired teachers were called back into harness, it was about time to return. Abandoned by Bertrand to our own devices, we had lost the habit of fantasizing. He looked at us with scorn.

"Nothing to say? Three hundred and forty-one miles in six days, maybe five if your calves are up to it. Adventure! And, at the end of the line, combat! I want to be there. What about you?"

"Us?" I managed to croak.

I was literally sweating, trapped by that "flying column" like a bourgeois by his principles. But I had to accept the challenge, even though it would fool no one, since everyone knew all along that once my back was to the wall, I would somehow wriggle out of it. Just thinking about it made me sick with shame.

"Not the girls, obviously," Bertrand answered. "The boys."

"At thirteen?" Pierrot objected with a brief shrug. "As soon as we get there they'll grab our ears and then ask for something to dry their hands with. All that way for nothing."

"We'll lie. Don't I look fifteen? Aunt Melly told me so." Then, turning to me, "What about you? You're big! You're even hairy!"

I acknowledged this with a nod, incapable of speech. Joking about my hairy feet was one of our conventions. Usually I laughed. Now I could have cried. The trap was closing. And

Bertrand had indeed aged in two weeks. There was something in his gaze, a more mature, more resolute expression, a determination no longer channeled into the imaginary.

Pierrot went on. "Even at fifteen, they'll need our parents' permission to take us."

"Do what I did, forge it. And of course, my father is an officer, he's far away, he may be dead, he can't sign anything at all."

A summary of the military condition . . .

"Your mother, then?"

"Exactly, my mother."

He shoved a bit of paper at us. On it were these words in his own thinly disguised handwriting:

To the Commander-in-Chief
Free Corps
  I give my son to France. Make a hero of him.

Laïcha
Oriental dancer

We were amazed.

"That's your mother's name?"

"Sometimes."

"And she's an Oriental dancer?"

"Sometimes."

He fiddled with his bracelet, and we got nothing more out of him. I have long wondered about the exact meaning of that note. Never, as I have said, did we talk about Bertrand's mother. We knew that he lived with her in an apartment in Versailles on the avenue de la Reine; it belonged to his father, the man who excelled at choosing distant, peril-fraught assignments. Aunt Sophie Majorel, Squadron Leader Carré's sister, veiled the whole subject with a sad silence, as though Bertrand had once lived with a socially unacceptable woman, now dead. I imagine that in "giving her son to France," Laïcha, "Oriental dancer," was making up for her condition, at least in Bertrand's mind: he had forged the note in a succinct and theatrical style of which he was clearly proud. I see no other explanation.

Folding the letter in four and putting it back in his pocket, Bertrand explained his plan.

"First of all, we'll go in twos. Zigomar, you're as blind as a bat. You couldn't tell a German from a Frenchman. You stay behind. As for you, Pierrot . . ."

Pierrot reacted with his customary candor.

"I'm not going either," he decided. "My father would skin me alive. Anyway, I'm all he's got."

Dear, crafty little peasant . . .

"All right, you don't go. But you and Zigomar have a very important part to play. We need two days' grace before they start looking for us. In two days we'll be far enough. They won't be able to catch us then."

We, we, we? Bertrand and me, of course! He hadn't even asked me what I thought. Knowing me well, he must have been certain. I have never really been sure whether he was giving me a chance by putting his trust in me, or else cornering me, knowing I would break. At the time, it must be recognized, my status went up in the eyes of the girls. I thought I detected in Maïté's eyes a spark of interest. Who knows? Perhaps some admiration as well . . . Pictures came to me of pants lowered in my honor, mine alone! I seized the golden moment, "They'll never get us! Count on me!"

Instantly, Maïté's gray eyes released me. A question of tone, no doubt. My words rang false. Nothing escaped her. I went red and affected a passionate interest in Bertrand's words. In fact, I groveled. Ah, it was so difficult for me to be a child among children . . .

Bertrand had unfolded his Michelin map. In red pencil, with carefully calculated dates and times marked in, even making allowances for the topography (in those days Michelin maps indicated slopes with little horizontal V's, single, double, or triple, depending on the gradient), the route ran from Châtillon-sur-Indre to Auxerre, via Romorantin and Aubigny, from Auxerre to Commercy, via Troyes and Saint-Dizier, and finally from Commercy to Forbach, via Pont-à-Mousson and Saint-Avold. At an average speed of seven miles an hour, we could reasonably expect to travel between fifty-five and sixty-five miles a day. The first two days we would need to move fast. First leg to Sologne, in the direction of Salbris. Second leg past Auxerre, past the village of Pontigny, to a forest Bertrand had marked in red. We would sleep under canvas, in woods, or in abandoned barns.

Brought down from its tree, the old iron trunk containing our "strategic reserve" provided our campaign supplies. Bertrand had not been idle these past two weeks: there were flashlights, tarpaulins, old horse-blankets, and a pair of 5.5 rifles, barrels unscrewed, oiled, wrapped with rags, pampered. We tested the bikes fully loaded. We stuffed six days' rations into our panniers, along with bedding and a change of clothing rolled up on the racks.

As I fastened the elastic cords with sweaty hands I forced myself not to think about what was going to happen. Bertrand had organized a full-scale delaying action. That very evening, with the aunts' permission, we would go camping at Pierrot's, in a field next to the farm, as we often did when the weather was fine. Not the girls, obviously, they weren't allowed to. They would join us in the morning for a last breakfast of candy and goose liver, in the dew under the rising sun. Then Bertrand and I would discreetly head off down a dirt road that skirted Petit-Bossay and pick up the road to Châtillon-sur-Indre farther on. With the help of Zazanne and Maïté, Zigomar and Pierrot would throw off suspicion. The camping would serve as a blind for two days. The farmhouse had no phone, and since the declaration of war we had enjoyed even greater freedom than usual, as if our aunts and last remaining uncles had delegated to us the responsibility of recovering what could be salvaged of the joie de vivre that had deserted all our lives. After the two-day delay, once our absence was discovered, those who remained behind would uphold their sworn silence, at least as far as our itinerary and destination were concerned.

And so it was that next morning, in dry, cool weather, the sun shining, divine, real hiking weather, I found myself dazed and wheel to wheel with Bertrand, pedaling with all my strength down the road to despair.

After two hours we stopped to quench our thirst at a village fountain. No one here knew us or wanted to know us. The church clock struck ten, and three disturbed pigeons flew up, only to alight again a moment later. An old woman went by, pushing a grass-filled barrow, her clogs clumping audibly. She didn't so much as look at us. A kid came and stood right in front of us, then ran off without asking questions. The peasant world has never been very sociable, but this was amazing. I now believe that the farmers were already beginning to back away, that they were turning inward, beginning their great five-year withdrawal to their farmyards in order to get through the bad times whose approach they atavistically scented, as animals smell fire. These details are unimportant, but they are the memories that came to me as I tossed through those feverish hours, along with the name of the village, Vatay; and today they seem quite remarkable. Yet I had methodically crushed them too into the back of my mind, twisting them out of shape. Oh, I do not like those memories . . .

We were sitting on the rim of the fountain.

"How many miles have we gone this morning?" I asked.

Bertrand checked the map.

"Twenty. Another thirty-six for Salbris. We'll be there before nightfall. OK with you?"

"I'm ready to go to the ends of the earth, but I'd just as soon get there quickly."

He gave me a chance to extricate myself gracefully.

"Why? Are you really that set on fighting? Look, I forced you a bit, but if you want, you can change your mind. You won't be dishonored."

I didn't hesitate for a second. I exploded with protests.

"Of course I want to fight! Why do you think I came? Did I back out, like Pierrot?"

He looked straight into my eyes. I don't know what he found there, perhaps a reflection of my determination just before it collapsed, because he held out his hand to me and smiled warmly.

"Perhaps I misjudged you. Until death us do part, you and me!"

Every word a dagger thrust into my heart; he probably didn't realize it, although I still wonder sometimes . . . There was a pool of calm water just below the fountain; it showed me my reflection, face grave, gaze manly beneath the Southerner's cap, jaw muscles tensed in token of my resolution: and all bluff! At that precise moment I was calculating that I would have to do those thirty-six miles all over again, and quickly, plus the twenty already traveled, for I knew that nothing in the world, not even the fires of my own contempt, would make me carry on past Salbris. Picking up a cobblestone, I threw it hard into the basin and my beautiful image disappeared. That was my only act of honesty that day.

"Shall we go?" Bertrand asked.

"Let's go!" I said.

And it was I who took the lead, with Bertrand in my wake!

When we reached Romorantin at around four in the afternoon we ran into a roadblock. The road was crawling with policemen. It was too late to turn around without giving ourselves away; we were under their very noses. Something must have gone wrong with our plan. It could only be us they were looking for. Praised be the Lord for returning me safe to the fold like this — not as a shamefaced coward but as a tragic hero.

"Let me handle them," said Bertrand. "I'll do the talking."

It wasn't necessary. They asked us nothing. They weren't even interested in us. Someone had heard an airplane over Sologne during the night. The police were looking for a German spy disguised as a Sister of Charity or a postman or a policeman, depending on the witness, and every policeman in a thirty-mile radius, carbine in hand, had been demanding other policemen's identity papers all day long. Nothing was too underhanded for these German spies. It was our luck that they hadn't thought of disguising themselves as children. We got through like a letter through the mail, a hot knife through butter. Pedaling on along the road to Salbris, Bertrand had wings. I had legs of lead.

In those days Sologne was still hospitable country. The road
we were traveling didn't bristle with spiked fences on either side
as it does today, there were no threatening signs to let the traveler
know that he is tolerated on the highway only because the
authorities can't do otherwise. We were therefore free to choose
from among any number of wooded byways that trickled down
toward Sauldre, at whose outskirts we made camp on the
grounds of a property fittingly named La Sauldre. We leaned our
bicycles against a tree by the river and swiftly pitched our tent —
one tarpaulin draped over two poles, another as a ground sheet.
Bertrand lit a fire between two stones on the bank. The country
seemed deserted, and the game, no longer hunted, cackled and
squawked with happiness. The air was light, sweet-smelling,
mild. Night could come, we were prepared.

We ate warmed cassoulet from a can, cookies, and chocolate.
We talked about our adventure, our plans. Or rather Bertrand
talked, and I (my heart tight in my chest, because every word he
uttered condemned me) had barely strength enough to keep the
ball rolling. At one point he said to me, "I've been bluffing. I'm
not sure they'll want us in the Free Corps, or anywhere else in
that old man's army of ours."

"What's all this for, then, all this traveling?"

"I'm not sure, but there's just a chance. We have to try. You
always have to try. And I want to have something to be proud of.
There were a few of them among the men who left — men who
were proud, who kept their chins up, whose eyes shone. If it
doesn't work out, if they send me back home" — he stood,
adjusting his cap, miming the scene — "I want at least to have
challenged the commander of the Free Corps, clicked my heels
at three paces, saluted, and straight off, as if we were in a military
staff meeting, yelled out, 'I am Bertrand Carré, only son of
Squadron Leader Carré of the foreign legion, posted to the
Chinese frontier, fifteen years old, crack shot, knife thrower,
acute nocturnal perspicacity (I looked it up, it means I can see at
night), volunteer for all the most dangerous missions, engaged to
the most beautiful girl in the world but as free as the air I breathe.

I have come to request the honor of enrollment in the Free
Corps, and the honor will be reciprocated!' That's got guts, no?"
    I had to agree. He was glorious, his chin sharp, his eyes the
blue of a steel blade, his tone inspiring. I admired him, I hated
him, I wanted to sink into the ground. Bearing plus attitude
equals conviction, I can't get away from that. I have always lacked
one or the other of those.
    "Good. Perhaps we should sleep now," he said.
    I rolled myself in my blanket fully dressed.
    A little later, with the flashlight off, in the darkness, he con-
fided, "Whatever happens, you'll always be my friend."
    His friend or his audience? Forty-seven years later, that ques-
tion is still unanswered.
    A moment after, he was dead to the world.
    I was exhausted in body and in spirit. But it was vital not to
surrender to sleep, or at least only in small doses, in order to
catch the first glimmerings of dawn and flee without waking
him. It was fear that kept my eyes open. The most horrible night
of my life. I relived it in my fever, like the nightmare that it was. I
cringed at the thousand sounds of night in the forest. I heard
them, listened for them, lay in wait for them, branches snapping
as though stepped on, night birds hooting, the voices of leaves
rustling in the wind, a whole universe of lunar creatures that
loomed hideous and thirsty for blood as I lay there in a half-
waking state. I roused myself with a start, bathed in sweat, and
found myself faced with myself. I was my own nightmare. At
around three in the morning a stag belled close by, and then
there was a terrifying clatter through the undergrowth. Desperately, I recited all the prayers I knew, including one I had
invented myself, with a wealth of variations so that it could be
reeled off like a litany: "O God, take pity on cowards for they are
the most wretched. O God, take pity on the craven for they are
the most wretched. O God, take pity on the chicken-hearted
. . ." The frightened, the timid, the yellow-bellied, the scared,
the terrified, the funkers, the fearful, the faint of heart, the
nervous, the timorous . . . God did not listen, but at least the
prayer peopled my terror. Time moves slowly when you are

thirteen and in an unfamiliar forest at night, planning a shameful deed for the morrow. At last a cock crowed. I don't know if he crowed the requisite three times, because it took me only a few seconds to jump on my bicycle and speed out into the new morning, with Bertrand still asleep behind me.

I have asked myself many questions about that flight. Why did I pretend I was Bertrand's trusty lieutenant—all the time knowing I would abandon him—when I could simply have refused to go in the first place? There was no lack of good excuses. And why had I refused his offer of an honorable release when he proposed it at the Vatay fountain, only to walk out on him a few hours later, shamefully, without a word, without an explanation, without a good-bye, without even meeting his eyes? Perhaps from some inverted sense of honor. If you had to be a coward, why not be one all the way, to the hilt? An ugly role, but at least a genuine one. Unbeatable in his own chosen role, Bertrand hadn't left anything for me. There perhaps lies the explanation . . .

By six that evening I was back—an hour ahead of the agreed "delaying" time, as well as the time limit the aunts had imposed on our camping. I made a wide detour around Pierrot's farm, trembling at the thought of coming upon one or the other of them, Maïté, especially Maïté . . . Everything seemed calm at La Celle. The axe had not yet fallen. Taking advantage of the sun's last few rays, Aunt Melly was daydreaming on her wicker lawn chair, an open book on her knees.

"Ah, there you are. Did you enjoy yourself?"

I mumbled some answer.

"But just look at you, you look exhausted! Is anything the matter?"

"Everything's fine, Aunt, I swear."

"Come here."

To admit everything, all at once, forthwith, to release the anguish that had accumulated within me until my very breathing hurt, then flee to my room, unburdened at last, and cry into my pillow until it was all out, the fatigue, the emotion . . . But I didn't have that kind of courage. No matter what I have tried, I have never once in my life been able to force fate's hand. I have always let it take its course.

"You really don't look well," Aunt Melly went on. "You children have been running wild lately. War or no war, you must start thinking about going back to school. Have you finished your summer homework?"

War or no war? In millions of French households, khaki sweaters were being knitted, and the army postal services daily ferried millions of packages stuffed with millions of sausages, pipes, and pairs of socks. Conscripted wine growers were being sent home for the grape harvest, and every evening the doughty Captain Lavallée, my uncle Sébastien, phoned Aunt Melly from his Loches command center, set up in The Three Ducks, a hotel renowned for the game dishes from its kitchens. Yes, I might as well finish my summer homework, forgotten since September 3.

At nightfall we sat down to dinner. At La Jouvenière, home of Aunt Sophie Majorel, they would just be beginning to worry over Bertrand's failure to return. My cousin Louis Lavallée, my elder by five years, fiddled with the knobs on the radio as the general staff made its daily oracular pronouncement: "Quiet all along the front. A few brisk skirmishes in the Warndt Forest." At the telephone's first ring I went as white as a sheet, my fork frozen halfway to my mouth. In those days, in the country, you did not simply pick up the receiver and immediately identify the caller. First there was a whole series of feminine voices, of clicks, of static. You waited.

"Hello, Petit-Bossay, this is Châtillon, darling. Here's Loches . . ."

"Hello, Loches, this is Petit-Bossay. I'm listening, darling . . ."

And Loches would reply, "Give me Number 1, darling, for Loches 124 . . ." That was Uncle Sébastien, from the phone box at The Three Ducks. He was fine. He was bored. He would come home next Saturday . . .

The axe fell when the phone rang again during dessert. This time it was quick, a local call: Number 4 in Petit-Bossay was calling Number 1. Number 4 was the Majorels. The wall phone was in the entrance hall. My aunt had left the door open. She listened in silence, and then I heard her answer, "What? No, no, my dear Sophie, he's here, I assure you! He came back just a little while ago. We're finishing dinner . . . Don't hang up. I'll ask him . . ."

She came and stood before me, looking more intrigued than angry.

"What's this I hear? Apparently you're supposed to be a two-day bicycle ride from here, pedaling God knows where with Bertrand Carré. He's not at home. Where were you going? Where is he headed? Whatever came over you? Your friends refuse to talk. The Majorels are terribly worried. They're responsible for that child."

I remained mute, head lowered, eyes staring straight ahead, utterly absorbed in evading a new moral dilemma. I had just realized that I hadn't sworn anything at all and that Bertrand hadn't asked me to, simply because he thought I was going with him. Except, of course, that here I was back again. But for the sake of honor, should I not keep my mouth shut, like the others?

"Well?" Aunt Melly insisted. "Lost your tongue?"

I said nothing.

"In that case we'll have to gather everyone together here and get to the bottom of this," she said. "There's no time to waste."

I immediately found my tongue. "Maïté as well?"

Maïté's gray eyes, full of disdain . . . I could already feel them on me.

"Oh, no, Aunt," I blurted out. "Just Uncle Armand and Aunt Sophie, please!"

"Then will you tell us where to find Bertrand?"

I nodded.

Aunt Melly went back to the telephone. Ten minutes later the Majorels had joined us. I knew the itinerary by heart; we had studied it enough! Aubigny, Auxerre, Troyes, Saint-Dizier, Commercy, Saint-Avold, Forbach . . .

"Forbach!" Uncle Armand exclaimed. "What were you going to do in Forbach?"

"Join the Free Corps," I answered unconvincingly, trying to throw out my chest and lift my chin. It didn't work: that was Bertrand's role, not mine.

I vividly remember that moment. I was expecting scorn, dismayed shrugs, angry remarks about our age and mental health. But Uncle Armand smiled warmly.

"I'd have done the same at your age," he said, "and even at my age if I could. But they don't need old men or boys for this war. As far as heart and courage are concerned, that is perhaps a pity . . . And whose idea was it?"

I was tempted to claim, "Both of ours together," and to hold a portion of Uncle Armand's good-natured sympathy to my wounds like a healing salve. But although I could lie to myself, I couldn't lie to this fat, kindly old man.

"Bertrand's," I said.

"Of course! Bertrand all over," he commented happily. "And you, my boy, don't look so glum! At least you tried, that's not so bad. Right! Now to track down the other hellion . . ."

We opened up the road map and with my guidance pinpointed him camping somewhere in the Toucy region, this side of Auxerre, and planning to reach Brienne-le-Château by next evening.

"Saint-Dizier, more likely," said Uncle Armand as he headed for the telephone. "He'd really be killing himself to get any farther. Hello, darling, give me the Châtillon police station . . ."

"Hello, Number 1, you have the police station . . ."

Uncle Armand took to the road early next morning at the wheel of his Celtaquatre and was back by nightfall, Bertrand sleeping like a log in the backseat of the car, his bicycle packed away in the trunk.

The police sergeant at Montier-en-Der, a town in the Haute-Marne district twelve miles past Brienne, had returned Bertrand to his uncle with these words: "Made of the right stuff, that kid! When I flagged him down he said, 'I'm glad you're here, Officer!' as though he was talking to a subordinate. 'I would like you to escort me at once to the commander-in-chief of the Free Corps in Forbach. I am Bertrand Carré, only son of Squadron Leader

Carré of the foreign legion, fifteen years old, crack shot, knife thrower, acute nocturnal perspicacity, volunteer for all dangerous missions . . .'"

I knew the rest. So he had managed to work his speech in after all! Perhaps he would hold less of a grudge against me?

Two days later we saw one another again, thanks to Uncle Armand and his Celtaquatre. Our bicycles had been confiscated. He didn't seem to hold my defection against me. He didn't say a word about it. He had already turned the page. Blue Island, the La Jouvenière bus, that was all over. We no longer had permission or the means to get about. We lived out the tail end of our extra vacation under house arrest. Then the Petit-Bossay school reopened and swallowed Zazanne and Pierrot. Maïté went back to her fashionable boarding school in Tours, Zigomar to a school in Blois. And Bertrand and I entered our third year, he as a boarder at the Saint-Geneviève School in Versailles, and I as a day student at Janson de Sailly, in the sixteenth arrondissement in Paris that was so dear to my parents.

I went back to a Paris at war, much like a Paris at peace except for a few small differences. A not insignificant segment of the male population was in uniform and seemed to have been mobilized solely to occupy sidewalk cafés. The city was covered with beige X's, strips of wrapping paper glued to windowpanes and to glass storefronts. Gas and electric streetlamps had been swathed in blue. The nights were pierced with the whistles of civil-defense block wardens, solid citizens who tried to look important but whom people chose to mock with inimitable Paris humor, attributing to them the kind of marital woes normally reserved in French lore for stationmasters. People strolled the streets, their gas masks (a precaution quickly dropped except by a handful of fearful dotards) dangling from shoulder straps and slapping at their buttocks. When meatless days were decreed, my parents dined on prawns and grilled sole, and *Le Petit Parisien,* which our chauffeur brought in every morning, suggested menus for distracted housewives based on calf sweetbread, liver, tongue, brains, blood-pudding, sausages, chicken, guinea fowl, chitterling sausages, ham . . . The war, I may add, had deserted the front page for page four, a few quick lines quickly disposed of, centering on the predictable staff communiqué: "Quiet all along the front." And it was for *this* that we had mobilized millions of soldiers! Who naturally enough weren't happy about it. So the generalissimo, ever fatherly, dreamed up an Armed Forces Entertainment Organization and brought Momo and his latest hits to the front: "Victory (*repeat*), the daughter of Madelon. Victory (*repeat*), it's our marching song. The nation can't be wrong. Victory, (*repeat*), the daughter of Madelon . . ."

Bertrand brought me the record one Sunday and played it on my phonograph, licking his chops over it before it even started.

"See? See?" he said. "We've won already!"

He came once or twice a month to spend his Sunday pass in Passy. I think he missed me. He wasn't a good student. A dreamer, gifted but lazy, he kept his books at a distance, as though with tongs, much as he did his classmates. His real audience was me. Once he shoved a column clipped from *Paris-Soir* into my face, entitled "Life at the Front." From it he read that in six months of war 128,207 soccer balls and 674,583 decks of cards had been sent to the front.

"So, what do you think?" he asked me.

"Should I think something?"

He was my friend. Why should I steal his dramatic thunder?

"This means," he said, "that we can place blind faith in our military leaders. At least they can count. With two million soldiers in arms, at twenty-two players for a game of soccer and four for a game of bowls, it works out, we've already won!"

The games went on peacefully until four in the morning of May 10, 1940.

I AM NOT writing a child's memoirs of war. Just the account of a short period of my life that might have had no more effect on me than the five dramatic years that followed, had Bertrand not been there to bring it all to life. In the course of my illness the fever pushing open the locked doors of my childhood operated selectively. It exhumed just that one episode: Bertrand. As for the rest, I disappear from my own memory as soon as his illuminating presence fades.

I therefore have no clear idea of how I was affected by the succession of national disasters whose main consequence was to bring the six of us together again on Blue Island in June 1940. There was the collapse at Sedan, the last stand around Dunkirk, the retreat, and then the rout that in twenty-eight days brought the Germans to the banks of the Marne, the Oise, and the Seine at the gates of Paris (Momo, of course, had been among the first to flee). My memory does not come alive again until the evening of June 6, 1940. Bertrand was at our apartment when my father called from his office to say he was sending a car; he wanted us to be ready, bags packed, within the next half-hour, to join an official convoy at the Invalides. It would be made up of the advance parties of every government ministry, its mission to pave the way for the probable evacuation of the government to Touraine. As soon as a final decision had been reached my father and his boss, Chabannais, would join us. He wanted my mother to help arrange accommodations for Chabannais and his staff at La Celle, Aunt Melly's home. The foreign ministry staff, led by Alexis Léger, would probably stay at La Guichardière with my aunt Octavie de Réfort.

"And Bertrand?" I asked. "Let's take Bertrand, Mama. He can stay with his aunt Sophie at La Jouvenière."

"But what about your mother, Bertrand?" objected mine.

Bertrand shrugged, picked up the phone, dialed his number in Versailles, explained the situation in ten curt words, then passed the receiver to my mother who listened in silence to the answer and hung up without saying a word.

"So . . ." she said thoughtfully, shrugging in her turn. "My poor child, we'll take you with us."

The "poor child" was overjoyed.

Which is how Bertrand and I found ourselves comfortably slumped in the padded foldaway seats of a ministerial limousine as roomy as Kolb's omnibus, attentive and curious as owls on a nocturnal expedition, across from my mother, who was carrying a dainty little valise on her lap, and a swarthy, rather unpleasant man loaded down with fat leather briefcases; we at once nicknamed him "Flat Top" for his round flat head. He cast venomous looks at Bertrand, who took pleasure in staring him down, and he relaxed his jaw muscles just once, to complain that he was hungry and thirsty and to wolf down a chocolate bar he offered to share with no one. Armored cars from the riot police escorted the convoy; and it was preceded by a motorcycle squad that cleared a path for us through a mass migration of apocalyptic dimensions. We had been asked to be discreet, to avoid showing ourselves. Curtains covered the rear windows. A government in flight doesn't excite much sympathy, and God knows what those fifty black limousines contained in the eyes of the unfortunates they pushed to the side of the road! It was always the same ones, the powerful and rich, who got away with murder!

At one point Bertrand and I peeped through the curtain. There, close enough to touch as the car inched past them, stood three pairs of exhausted horses harnessed to a gun carriage. The drivers swayed sleepily on their seat. They had been trapped in the middle of an endless river of people and vehicles, civilian and military all jumbled together, its headwaters in Paris but its course swollen by muddy tributaries from Normandy, Champagne, and Picardy, all flowing sluggishly toward the bridges of the Loire, and they had given up the idea of taking part in a battle sputtering out far behind them and without them. They had decent faces. Bertrand gave them a brief salute — in circumstances like those a gesture more likely to invite insults than blessings, but his charm worked wonders. They briefly came alive, responding with a weary smile that seemed to say that even if our goose was cooked it didn't rule out friendship, particularly since we would never see each other again.

"You see," Bertrand whispered in my ear, "they agree with me: we've won."

"Will you close that curtain!" my mother said. "Show a little tact! You almost seem to be enjoying this."

The intoxication of disaster . . . It terrifies young children, but it grips adolescents with a kind of exultation. Time speeds up, old habits disintegrate, codes of behavior collapse, the unexpected constantly occurs, masks fall from faces, the adult world goes head over heels . . . all of which adolescents find prodigiously entertaining.

At one point the convoy came to a halt. It must have been a little before Blois. Army motorcyclists drove up alongside the column, shouting to our chauffeurs: "Move over! Get to the side of the road! Relief troops coming through! Move over!"

Flat Top opened the door and barked at one of the escort officers, "What's this? You've lost your mind, Captain! We have absolute priority! Pri-o-ri-ty, d'you hear?"

Manners no longer mattered. Time was flying. June nights are short. Daylight might bring Stukas interested in taking a closer look at our somber, conspicuous convoy. From every limousine a high-ranking official emerged to trumpet his claim to priority. Women with dyed hair poked their heads through the windows. Secretaries? Mistresses? Both?

One of them called out in a throaty voice that made me think of Zazanne, "Do something, darling! Are you the boss or aren't you?"

My mother made a helpless gesture intended for Bertrand and me, as though she wanted to apologize for inflicting such company on us. On the other side of the road the endless stream of those forced to flee on foot had also come to a standstill: people from eastern and northern France, silent, not undignified (Paris had not yet begun to disgorge its masses). Disbanded troops without leaders or orders. Plus officers in private cars. Nothing exciting, just a nation in defeat. The exodus has been described a thousand times. I will add nothing, except to mention the disgust on the face of an old peasant from Lorraine who sat perched on the piles of belongings crammed into his wagon and watched Flat Top, close to hysteria, flap his arms.

"Priority!" he was screaming. "We are the government . . . make these cartloads of yokels stand aside for us instead!"

The officer in command of the troops, who had asked us to pull over, inhaled deeply. Then, politely, not raising his voice, pronouncing each word separately and distinctly, he replied, "Fuck yourself, Mr. Bureaucrat."

Shaken to the core, Flat Top jumped back into the car, forgetting even to close his mouth, and Bertrand, who had also stepped out onto the road, held the door open for him with an insolent smile and the exaggerated gestures of a demented footman.

"After you, Mr. Bureaucrat," he said in clear tones, mimicking the officer's manner.

We heard the sound of engines at full throttle coming up from the south. A motley steel squadron. A tank with the name "Kléber" painted on its turret and the green pennant of a cavalry regiment fluttering from its antenna; a second tank, baptized

"Marceau," two open armored cars, their eight machine gunners upright on their seats in fours, face to face and as rigid as if on parade; then a small tank truck . . . and nothing more. The relief force had come and gone. The tank crews' leather-helmeted heads emerged from beneath the raised hatches of gun emplacements and loopholes — twenty-year-old faces, bright eyes staring straight ahead, oblivious to the ragged mob milling on their path, a mob they could easily have crushed beneath their treads, probably with no regrets. The second-lieutenant commanding them, half his body rearing centaurlike from his tank, instinctively spotted and saluted with hand and eye a boy with tears in his eyes standing by the side of the road, watching this mirror of himself go by, this projection of his own youth. It was Bertrand.

All heads were now turned to the south, watching for the bulk of the troops heralded by this vanguard. Nothing more came. I believe that everyone in that headlong retreat who was wearing a uniform, and had turned his back on the fighting, now stepped out again with light heart and confident steps. You don't win back a war already lost with the two dozen rash whippersnappers we had just seen go by, thinking they were setting an example with their fire-eaters' faces. When it's over, it's over, when the goose is cooked, it's cooked . . .

"We went through all that trouble for *that?*" Flat Top sneered. "We lost precious time for *that?*"

Bertrand jumped in front of him, thrust an accusing finger toward him, then extended the gesture to include our whole convoy of ministerial limousines, flying official French flags the better to forge through the mob.

"And it's for *that,* Mr. Bureaucrat," he said, "that they're going to fight?" Then the finger swerved back to Flat Top's comfortable belly. "And it's for *that* they'll be giving their lives? I hope, for their sakes, Mr. Bureaucrat, they'll find a better reason than *that!*"

"Did you hear that? Did you hear that?" said Flat Top to my mother, choking with rage.

"I heard," my mother answered, her voice expressionless.

"Oh, really . . . Oh, really . . ." he gulped. "I . . . I . . ."

There was no follow-through. I concluded from this that in Chabannais's eyes my father must be more important than Flat Top.

"Bull's-eye!" I said, sure of my ground.

"Please," my mother said softly. "Not you."

But she smiled at me affectionately.

And the flow rolled on once more. A whole people flowing slowly south, four-wheeled peasant wagons, baby carriages, bicycles, two-seater bikes from industrial suburbs, tiny cars sagging beneath the mattresses lashed to their roofs, and a whole army of exhausted footsloggers, old men condemned to walk, squalling children in their mothers' arms, we'll never see the likes of this again, will we . . . who knows? Moving much more rapidly, leaving its faithful constituents in the dust of its motorbike escort, His Republican Majesty turned his back on the truth and fled in the direction of his châteaux.

For the sake of accuracy I have even checked the facts to see if my memories tally with reality. It was indeed on that night of June 6, at ten-thirty, that General Weygand, the commander-in-chief, told an astonished war cabinet on the rue Saint-Dominique that he had just brought up his last armored regiment, cobbled together in Blois from various training units, and that this was his last reserve force. It was this "regiment" I had seen go past—Kléber and Marceau . . . And on that same night, at the same meeting of the war cabinet presided over by Paul Reynaud, Marshal Philippe Pétain, France's newly appointed deputy prime minister, first uttered the fatal word "armistice."

After Blois, the government cortege split up into twenty or so miniature convoys of three or four limousines each, ministry by ministry, to disperse within a radius of thirty miles around Tours, both north and south of the Loire. At around two that morning, each group felt its way through the hollows of dark valleys or between stands of tall trees for the particular château it had been assigned, half of whose owners had not even been advised of the impending arrivals. At La Guichardière my aunt

Octavie de Réfort was dragged from her bed at dawn by a tall and somewhat self-important visitor in a wide-brimmed fedora, pallid in the early morning light and surrounded by a swirling bevy of aides and attachés, nervous as ticks, and by more robustly built chauffeurs dumping enormous expensive suitcases on the front steps. He introduced himself as Alexis Saint-Léger Léger, general secretary to the minister for foreign affairs, and announced that he was there (without feeling the slightest need to apologize) to set up office in the requisitioned château. Family legend has it that Aunt Octavie, wearing a dressing gown buttoned to the throat, looked the poet of the *Eulogies* and of *Anabasis* up and down and answered in her most upper-crust voice — and God knows she could be a snob, much more so than Alexis Léger: "I am sure you have your reasons, sir, but this is nevertheless rather *thoughtless* on your part . . ." (The word she used, *thoughtless,* is "léger" in French.) Not a very subtle reply, thanks perhaps to the early hour, the rude awakening, and her visitor's unabashed vanity — but one that has made it impossible for me ever to read Saint John Perse without laughing.*

It was even earlier when we reached La Celle. My mother knew the way and guided the chauffeur through the web of country roads. Aunt Melly was waiting for us. She hadn't gone to bed. She was perfect. She kissed my mother, her elder sister, most tenderly, and they exchanged a torrent of family news. She asked my mother what she would have for breakfast. Then she gave a sharp reply to Flat Top, who was showing signs of impatience and demanding a bath and a telephone: only the minister himself would get a bathroom, she told him — her house was not a hotel for fleeing bureaucrats — and he and "all these other people" would have to be satisfied with second-floor bedrooms well provided with chamber pots; and furthermore, so that Petit-Bossay's assistant postmistress should get her full night's sleep, phone service was disconnected from ten at night to six in the morning. Only then, with a suspicious tremble in her

---

*Saint John Perse is the pseudonym of famous French poet Alexis Saint-Léger Léger. — TRANS.

voice, did she express her compassion for France, reduced to entrusting the defense of Loches to a man as peaceable and as incompetent as her "poor husband," Uncle Sébastien. Then, noticing our presence, she first kissed me, saying a few affectionate words, and turned to Bertrand.

"Well! You too, Bertrand. You've grown since last year. You look very handsome. Quite magnificent!"

Bertrand contented himself with the reply, "Thank you, Madam," accepting the compliment with a nod at once formal and comical, and a dazzling half-smile. I think it pleased him. Don Juan was no longer an older brother. Coming from my aunt Melly, the youngest of my aunts, this compliment amounted to a young male's initiation rites.

I do not want to give the impression that he derived inordinate pleasure from his looks. But he could hardly help noticing how often he captivated people with his fine, tanned, slightly feminine face, his blue eyes and direct, luminous gaze, his lithe young athlete's body, which always seemed to be draped in a Greek tunic no matter what clothes he was wearing. Many men and boys of his age hated him on sight, while others took him and his good looks tolerantly in stride. Women simply marveled, openly rejoicing that male attractiveness could scale such heights of perfection. Aunt Melly was no exception. I remember her as a pretty woman with a sharp mind, lighthearted, gay, who attracted looks from the young, knew it, and laughed. She would always make summer pets out of a few of the cousins, the nephews, and their friends; and no one ever resented it. It was a game. The brightest of these overgrown adolescents formed a sort of court around her, sitting at her feet in the garden, kissing her hand somewhat lingeringly in the soft spot between palm and wrist, and in exchange she would occasionally lay a quick hand on one's cheek, or ask for an arm to take a walk in the garden. It was of no real consequence. Custom had not yet imposed today's hideous blurring of age and gender distinctions. In any case, those games were over. War had put an end to them. Moreover, looking back, I am inclined to believe that it was out of habit, a memory of coquetry, that she distractedly brushed Bertrand's cheek with her fingertips.

"What will you come up with this time?"

And then she thought no more of it.

He came up with something, as we will see.

The next day, at six in the morning, Petit-Bossay's assistant postmistress duly reconnected the telephone service. As not a single one of the ladies and gentlemen from the ministries was awake, neither Flat Top at La Celle nor Alexis Léger at La Guichardière, the aunts dashed for the telephone.

"Maïté will jump for joy!" her mother, Aunt Octavie, said when she heard of Bertrand's arrival. This was of course a manner of speaking, for even at the age of five Maïté's self-control had been total; even then she could freeze your insides with her gray eyes.

At the wheel of her old Celtaquatre, Aunt Sophie came to take Bertrand to La Jouvenière. Having her beloved nephew back had filled her with energy.

"Magloire will soon be running out of gas," she announced almost gaily. "I'll be going back to pedaling, like the children!"

We agreed to meet on Blue Island for a swim as soon as the bicycles had been brought down from the attic: Maïté, Zazanne, Pierrot, Zigomar, Bertrand, and me, all henceforth free as the wind, schools closed, permanent holidays . . .

Flat Top clucked like a hen when he learned that an air force engine repair shop, pulled back from some distant air base knocked out of commission in the first days of the German attack, had been set up under canvas in the woods not far from La Celle, and along with it a single, ridiculous, utterly outdated anti-aircraft gun, now under camouflage netting on the fringes of the wood. He stamped in frustration.

"Insanity! What can the military authorities be thinking? It's just begging to be bombed! Have no fear, madam, I'll have that cleared out of here in short order!"

"But I assure you I have no fear, sir," said my aunt in her sweetest tones.

"But the minister! You don't expose a minister like this!"

"If the government is really leaving Paris, rest assured, it will merely be passing through."

On that nasty note she left to supervise the household: alternately skeptical and angry, facetious and on the verge of tears, but doing what she had to do. Tables had been lined up in the main living room; behind them idle speechwriters and typists, the passengers of our three limousines, sat enthroned. There were reams of paper, inkpads, dossiers. The women filed their nails. The men wore appropriate expressions as they exchanged lofty speculation about a situation beyond their understanding. The radio lied. The government lied. The phone rang without respite: "Hello, Petit-Bossay 1? Hold on for Charny 6," or "Chemille 4," or else "Cangé 17," or "Beaumont-la-Ronce 3" — all those seedy châteaux lost among the forests of the Indre-et-Loire region, all those ministerial advance posts, plus the prime minister's office in Cangé, desperately trying to contact one another through the offices of our local telephone operators. At noon the ministry personnel would get up and disperse for lunch, returning at two after a postprandial stroll.

Once I went to see the aircraft mechanics among the tall trees. Some of them were asleep. Others were playing cards. Two fellows in greasy overalls were tinkering listlessly with a single engine slung like a cooking pot from an iron tripod. What was the good? They didn't have the parts, they didn't know where their base had gone, and anyway there were no more aircraft.

In front of the main stairway of the château, the chauffeurs waxed their three black limousines, which gleamed like hearses.

We were at war!

THE CENTER OF the earthquake was coming close. Rather than trust the radio, which reported fighting on the Somme when the Germans had already crossed the Oise and "delaying action" along the Oise as they stormed the Marne and the Seine, we gauged the astonishing acceleration of events by the succession of shock waves that ran through our little world. The main road (the highway between Loches and La Roche-Posay), which had so far ferried a trickle of refugees along its provincial course, suddenly swelled to flash-flood proportions as Paris emptied. Even at Petit-Bossay, isolated on its backcountry road, June 7 saw the arrival of the first mattress-topped cars, which skidded to a halt when they spotted Magloire's pump. The aunts no longer knew which way to turn. My aunt Germaine Bonnadieu, at Beausoleil, found her courtyard blocked by six moving vans loaded with paintings from the Louvre; their drivers slept, exhausted, heads resting on their steering wheels. At the Durands', La Cornetterie, Zigomar's mother had to face fifty gaunt inmates of an insane asylum; they were sedated to the eyeballs and had shuffled all the way from Compiègne in hospital slippers behind a horse-drawn cart piled high with canned goods and treatment records; they were under the care of a single orderly, the others having abandoned them along the way. We bedded them down on the grounds. In the morning they were gone. In their place were twenty lost motorists from the pensions department of the postal ministry; according to Zigomar, they seemed to think they were on vacation and proved less well-mannered and much more demanding than the asylum inmates. But the height of terror came on the night of June 7, when, at the rising of the moon, the wooded grounds of La Jouvenière were filled with horrific roars. It was the Médrano Circus, also headed south, camping out in the countryside because no village would take them in. The lions had eaten nothing since the day before. They could be heard hurling themselves against the bars of their cages, and Kolb, his blunderbuss under his arm, rushed along the garden paths, believing all night long that he was hunting lions.

To return to the Germans, the armored divisions of the Reich were always two or three days in advance of our own communiqués (if I can judge by the history books I consulted in order to put my memories into chronological order), moving at such speed that villages supposedly a safe distance behind the front lines suddenly found themselves still far from the front and just as safe — but on the German side. The lightning war had come and gone. That is also what Lieutenant Frantz von Pikkendorff of the Fifth Panzers (an armored corps under General Hoth, engaged above the Seine, near Meulan) noted in his diary. I will explain later how this diary came into my possession. Here is what Frantz wrote on that same night of June 7:

Once it is dark the crews all raise hatches and keep their heads in the open for better visibility. There is no risk attached, for it has been three days since a shot was fired at us. We move at twenty-five miles an hour. The din we make going through villages rouses people from their sleep and brings them dashing into the streets, wearing slippers and waving for joy. They think we are the English! I wonder what perception of the situation they possess — or are being fed — to be sleeping peacefully on this night, as though nothing were happening. Then, realizing their mistake, they watch us for a moment and go back home to bed.

Next day, halted on high ground above the Seine:

The sappers rapidly repaired the Meulan bridge, as the French had blown up only one section during their retreat. Haste? Panic? Lack of orders? Botched improvisation? Perhaps the officer in charge of the task deliberately delayed the job out of pity, to give a last chance of escape to the leaderless rabble the French army has turned into since early June. I would rather believe that. Pity isn't a warrior's sentiment, but at least it is honorable. There is no longer anyone on the opposite bank. Our victory is total, devastating. As a German officer I am delighted. The humiliation of 1918 is avenged. To be living all this at twenty! I should be on a cloud, bursting with joy . . . I study my tank crews. Their features are creased with fatigue and lack of sleep, but their eyes are shining with happiness, with pride. They too are twenty years old, and it requires an effort of will not to betray the sadness I now feel. It is impossible for me to forget, seeing France on her knees, that I am half French as much as I am half German. I can imagine the suffering my mother would be going through today. I thank God for sparing her that. I shed many tears for her when she died two years ago, but now I realize that her death was a blessing. What could I have told her, the first time I went on leave, if she had lived? Unless I lied, what judgment could I make on her country that wouldn't have broken her heart? And my own . . .

We drink beer on a bistro terrace, under the plane trees of a charming little square where I halt my tanks to await the order to push across the bridge. The waiter serves us without resentment. I see window curtains parting, then people leaving their homes, driven by curiosity. They don't go so far as to speak to us, but they don't seem particularly grief-stricken. They are alive, their houses intact, their harvests untouched, the sun is shining, and we don't eat little children. The radio on the bistro counter announces that General

Weygand, the French commander-in-chief, is about to make a statement. The waiter casts a worried look at me and starts to change the station, but I ask him in French to let me hear it. I write down Weygand's words:

> The German offensive is now under way along the whole front, from the sea to Montmédy. Tomorrow it will extend as far as Switzerland . . .

I give a start. Switzerland! Guderian's Panzers must have flown like the wind on the left wing. Kleist, one of my tank commanders, who speaks a little French, slaps his knee at this news. He translates. The others guffaw. I ask them to be silent, which is unjust. They have every right to show their feelings. Poor France . . . Here is the rest of it, surreal in the extreme:

> Every French soldier's duty is to stand his ground without thought of retreat, eyes unwaveringly to the fore. Officers, NCOs, soldiers: your country's salvation depends not only on your courage but on all the toughness, the initiative, and the fighting spirit you can muster. The enemy will soon have shot his bolt. This is the last quarter hour. Stand firm!

I read incredulity and amazement on my companions' faces. For three days we have been forging through conquered territory without meeting any resistance. It is as if there were two quite unrelated wars going on, Germany's and France's. Where on earth have the French disappeared to? There is not the smallest sign of an organized body waiting for us across the Seine. In the field nearby, cows are grazing. This is it? The last quarter hour? An enemy who has shot his bolt? Fighting without thought of retreat? I watch my dozen rogues intently. I know them. I sense the pent-up mirth already shaking their shoulders. Kleist, speaking French in an accent that could be cut with a knife, does a passable parody of Weygand: "Ach! Stent virm! Stent virm!" and the laughter booms forth. I let ten seconds pass. They have a right to that. Ten seconds that are torture to me. Then I shout, "That's enough!" I put on my parade-ground face. I bring them to attention. They look at me curiously. They know nothing about my family, and I owe them no confidences, but I do owe them an explanation. "By belittling a vanquished enemy you belittle yourselves," I tell them. "You do not have the right to despise" (by which I mean: I forbid you to despise *me*, even if it is only half of me you despise). "The war isn't over, as far as I know. Perhaps you'll get the chance to admire the enemy instead. I hope you do" (which means: I hope I do). "The break is over . . ." And it is. Ludolf, my other tank commander, on duty at the radio, has just told me that we have received the order to cross the Seine. It is eleven on the morning of June 8 . . .

This diary was written in French. Not a single Teutonic turn of phrase. Frantz was by no means translating his thoughts word by word. He thought directly and effortlessly in French.

Weygand's order of the day was probably heard by millions of Frenchmen. I don't know whether they swallowed it like one of Momo's songs. So the fact that we heard it too, in Aunt Melly's living room at La Celle, where the six of us had met before going to Blue Island (each of us had raided his home kitchen for a picnic lunch), might be considered a natural coincidence. Today I know it was nothing of the kind.

Early that morning Bertrand and I, along with Maïté, had stationed ourselves at the bottom of the grounds of La Guichardière, behind a stone balustrade overlooking the main road where we could watch without being seen. There were more and more soldiers among the refugees pouring past, many of them on foot, black with filth, boots worn out, some of them dragging their rifles. And cars, vast numbers of cars, Paris buses, trucks from the Galeries Lafayette loaded with shattered army and navy personnel, a whole procession of black and red G7 taxis, at least six officers' caps to a taxi, caps with stripes and oak leaves, caps of every color, purple, black, red, blue, green. Bertrand, well versed in such matters (not a squadron leader's son for nothing), made sense of the hodgepodge for us: "Artillery, African Infantry, general in the Medical Corps, dragoon, Pharmaceutical Corps, quartermaster, frigate captain—what's he doing there? More doctors—but I don't see anyone sick—four generals, two from the Air Force, three colonels . . . well, well, what a mixture!" At first we thought it was fun. A kind of cruel game, a slapstick comedy that hadn't yet turned sour. We would point at random at people in the crowd. I remember a little gray soldier, leaning on a stick, limping along with an old man's gait, one foot in a shoe, the other in a slipper, and wearing a checkered cloth cap. Or the monocled colonel, preening like a conqueror, ensconced in the backseat of his requisitioned G7 as though he were headed for the opera, a young woman sitting next to him in a military nurse's uniform, waving her hand in dismissal, as though at flies, at the poor exhausted soldiers peering imploringly at the unoccupied seats of the taxi. Insults burst forth: "Slime! Bastard! Swine! He saves his tart and lets the soldiers die!" The colonel unholstered his revolver and thrust its barrel under the nose of a ragged sergeant clinging to the running board, and then the flow of traffic carried the scene farther away and we never learned its outcome. Bertrand smiled a nasty smile, once again taking up his sarcastic refrain: "Didn't I tell you we had won!" There weren't even any "relief troops" anymore, as there had been the day before on the Blois road, to blow a fresh breeze across this orgy of panic, this rout . . .

There are no half-measures when you assess your elders at the age of fourteen. For us, that crowd on the main road was the entire French army in flight, with France itself into the bargain. And yet the front was still quite far off. It would take the Germans at least another ten days to cross the Loire, then the Indre, the Mulsanne, and the Creuse, for they had chosen to concentrate the brunt of their efforts on the two wings, in the Brest-Bordeaux region and at Lyon and the Swiss border. Here again I have been able to confirm my recollections by leafing through the history books. That rabble represented army camps overflowing with listless, leaderless soldiers, desk units, the staffs of dozens of redundant headquarters, Army Service Corps depots, air force ground units, middle-aged stragglers from pioneer regiments, commissary personnel, anti-aircraft units, mechanics, orderlies, military police, hundreds of thousands of noncombatants who had no intention of being drawn into combat. They had all set off, without orders, as soon as the first droves of disbanded soldiers appeared on the roads, broken, not even defeated but swept away, and had mingled with them to form the pitiable herd streaming past below us. But it was not the French army. What was left of the army, although we did not know it, was being chopped to mincemeat somewhere north of Paris — gallant isolated units without leaders or orders, the young lieutenant on the Blois road and his tanks, Kléber and Marceau, most certainly among them, dug in around villages and refusing to pull back — while *180 miles to the south* and ten days ahead of the Germans on that morning of June 8 the remnants of what we took for the army (when in fact there was no more army) shuffled abjectly past below the grounds of La Guichardière. A sense of disgust, at fourteen. A lump impossible to digest.

At eleven a.m. that same day we gathered around Aunt Melly's radio at La Celle and heard the verbal heroism of Weygand's order of the day (". . . without thought of retreat . . . the last quarter hour. Stand firm."). Having seen what we had just seen on the main road, Maïté, Bertrand, and I, each in our own different way, reacted with fierce disbelief.

"Stand firm, stand firm!" said Bertrand, imitating the speaker's bombastic tones, swollen with the orotundity then considered de rigueur in radio broadcasting. "We're standing, general, sir! We're standing! We're standing on our heads, general, sir!"

"Bertrand!" Aunt Melly protested. "After all, people are dying!"

Sweeping the ground with an imaginary hat, he paused, theatrically and insolently.

"That is true, Aunt, I haven't forgotten it. Those who are about to die salute you."

"Bertrand! No more of that kind of joke, please. It isn't funny. It isn't decent."

As for me, I missed my cue as usual, striking an utterly wrong note. Even though (and I shouldn't have forgotten it so quickly) I too was of the breed that ran away . . . Seven years older and in uniform — and I would have been a part of that retreating rabble, hundreds of miles from Kléber and Marceau. At my witless refrain of "Stand firm! Stand firm!" Bertrand shot me through with a look, then said to me with a friendly poke in the ribs, "We'll stand, yellow belly, I promise!"

If that scene comes back to me so clearly, it is because it exactly parallels the "Stand firm!" episode described in Pikkendorff's diary. Two destinies on collision course . . .

That left Maïté. Her reaction staggered us.

"It's enough to put you off men," she announced blithely. "I like war. Win or lose, I would gladly have been the spoils of war. That's all I'd ask. But all those cowards down there, all those washouts . . ."

And Mlle Maïté de Réfort dismissed them from her exquisite person with a disdainful wave. With absolutely no effort of memory I see her again, Maïté transformed into a woman. And I acknowledge the happiness this rediscovered vision brings me. Breasts just slightly more swollen, hips and shoulders whose angles had merely softened a little, the same slender grace, youthful but assertive — yes, that's the word: a natural assertion of beauty. With bare arms and legs and with her long blonde mane let down, she had already reached her age of glory that spring. Some women reach it at twenty, others at thirty or forty, but the summit is within reach just once, and many of them do not notice it and weep for the rest of their lives at having failed to see it. Remembering the brilliance of her gray eyes that day, I now understand many things. She would not be among the weepers.

Mouths open, Zigomar, Pierrot, and I contemplated this new incarnation of the Goddess. Fat Zazanne (who always caught on long before us in these matters) turned red as a beet, and her whole body, its curves generously amplified since the previous summer, shuddered deliciously — and we three boys realized that behind the mystery, the rite of the lowered underpants, was something new to be added to the chapter on outmoded tribal customs.

Aunt Melly broke the spell. She looked at Maïté thoughtfully. "How old are you?" she asked in a soft, slow voice, like a card player making a wary opening bid.

"I turned fourteen two months ago, Aunt."

"And . . . ?"

The question, left suspended in midair, was understood. Pierrot, Zigomar, and I had no idea what was going on, but Zazanne turned scarlet all over again.

"Yes, Aunt."

"Do you realize that I should forbid you to speak as you have just done, that I should tell you to go home and advise your mother to lock you in?"

"Why, Aunt?"

"Do you understand the meaning of words – at least those that slipped out of your mouth?"

"They didn't slip out. I never say anything I don't mean and I always know what I am saying."

I can't remember what I felt at the time. A troubled surprise, perhaps. All this was taking place on the borderlands of that shifting, pitiless female landscape of whose existence I was still unaware and where I so often lost my way in later years.

"Well, I never! What self-assurance!" said Aunt Melly, affecting a mocking admiration. "Well then, since you think what you say and you know what it means, do you think it's appropriate at your age?"

"Definitely not, Aunt," Maïté went on in the same tones, "but war has changed all that. You have eyes and ears. What you have seen and heard in the past few days, is it appropriate? I saw an elegant young woman by the side of the road just a few moments ago, standing by a big car that had broken down. She had raised her skirt to her chin and was showing everything she had. She was shouting, 'Who wants me? I don't want to be left here. I'll give myself to anyone who wants me!' A carload of officers picked her up. It doesn't matter whether it was appropriate or not. What matters is that it was ugly, sad, cowardly, and despicable."

"I agree," said Aunt Melly. "But what to do about it?"

"The same thing, but with grace."

"You could at least wait. Nothing is forcing you – "

"Yes," Maïté cut in, "something is. War."

My aunt shrugged. "Words! You're just playing."

The answer burst forth, unexpected, "Perhaps, but you've been playing with us!"

Which left Aunt Melly thoughtful.

"That is true," she admitted, "and I shouldn't have allowed this conversation, but all the same . . ."

The woman and the girl exchanged a smile, a silent, mysterious token of initiation.

"All the same," my aunt concluded, "don't overdo it . . ."

It sounded very much like acquiescence.

LEAVING CHILDHOOD IS like climbing over a wall. You struggle to the top as best you can. You poke your head over and see a new landscape. And because there is nothing else to do but jump, you jump! You land as best you can. Some hurt themselves and take time to heal. Indeed, such injury can be a kind of premonitory death, a foreshadowing of the other death, of real death, the death to which a lifetime will inevitably — and all too often swiftly — lead you. It was our good luck to leave childhood by jumping into the middle of a real war. We found ourselves exactly as we had been before, but now playing in earnest.

From June 8 to June 10, 1940, for three days, Blue Island rang with the sound of axe blows and the rasping of saws. Bertrand had decided to "fortify" the island.

"So, no flying column this time?" Pierrot asked sardonically, giving me a distinctly unkind sideways look.

Renewing our pact of friendship from the year before, Bertrand came to my rescue.

"Leave him alone, will you!" he said to Pierrot. "You wriggled out of it too. We're a year older now. No one will back down. As for the flying column, this time it's German. And it's flying so fast we'll only just be ready in time. We'll wait for them here!"

He pointed at the bigger of the two iron bridges, the northern one; beneath it, the swollen Mulsanne tangled and disentangled its green waters like long coils of hair.

"Here?" asked Zigomar. "But no one ever comes by here, you know that. These Germans of yours must have maps. They seem to know where they're going. What makes you think they'll stray in this direction?"

"Because I am here!" Bertrand replied with unanswerable arrogance.

The bracelet glittered at his wrist. Like a grown woman, Maïté had slipped her arm under his. I have a clear memory of that scene. What a pairing of beauties! A rare couple, a regal couple . . . And let me now be very clear: you had to love them *together*. The carnal devotion of our ancestors to their sovereigns must have been something like that . . . Zazanne sniffled with emotion. She had grabbed my arm and was clawing it so hard she scratched me. Zigomar was beaming, in a seventh heaven. Pierrot tried to hide his adoration behind his stubborn peasant mask. Finally each of us returned to earth.

Like a good prince, Bertrand corrected himself, "Because we are here, you and I. This is our country, the six of us. The rest" — he gestured disdainfully, consigning all of France to the scrap heap — "we leave to them. To work! Let's go!"

We began by clearing the "lines of communication" we had cut the summer before through bushes and thickets all along the riverside on the island's northern bank. The result was a kind of sentinel's walk sheltered beneath an arch of greenery which allowed you to move about without being seen from the opposite bank. Drawing on his military vocabulary, Bertrand had divided the island into "sectors" with their own "firebases." We each — the boys — had one, a closed-in shelter, like a nest, from which our rifles could pour what Bertrand called a "crossfire" at the iron bridge. On our first firing drill, aiming at empty tin cans

planted on sticks as high as a man, we heard a few faint pings as
bullets struck home, but not one of the cans was pierced or even
knocked off its stick. Bertrand frowned.

"What did you expect?" asked Pierrot. "To stop the Germans
with 5.5's?"

"We'll stop them because we're here, that's all!"

Obviously. Everyone approved. Such is the spirit of childhood,
and Pierrot, who had briefly left it, was glad to scuttle back. If I
examine my memories closely, I don't think that we were really
fooled; in fact, Pierrot's remark proves it. And Bertrand most
definitely less so than the others. But to judge things from an
adult perspective—as we might well have done, being at the
frontier between two ages—it would no more have occurred to
us to question him than to sink in our turn into the general
helplessness that had seized our country. That Bertrand refused.
We followed him. No one would deprive him of his game.

This was even more obvious when he devised "reinforcements," fake soldiers made of sticks wearing poilu helmets from
the First World War—part of Uncle Armand's collection, now
prudently relegated to the barn in case of a German victory—and
rigged out in old greatcoats, hunting coats, ragged capes, and
brandishing antique weapons, rust-choked hunting rifles faithful Kolb had unearthed on the grounds of La Jouvenière. We
made eight of them in all, quite an army. We deployed them
along the "front," apparently seeking cover behind trees but in
fact quite visible. From a distance, putting ourselves in the
enemy's place on the facing embankment (but seeing with our
own eyes and with a will to believe that now strikes me as
pathetic), they looked strikingly true to life.

"They won't run away," said Bertrand. "They'll stand until
they're killed."

We were incredulous. And a little worried. A sneaking
apprehension. A word had made its entrance: killed. There is a
tone for games and a tone for reality, a tone for the imaginary and
a tone for the concrete. As in theater and in life. When parts are
well played, the difference is imperceptible, but a child's ear is
never fooled. Bertrand was mixing his genres . . .

While waiting we "held" the island, as Bertrand called it, in "staggered-defense positions," twenty or thirty yards apart along the covered path, each of us in his "firebase," his leaf-carpeted bunker of greenery. We practiced remaining utterly silent, scanning the opposite bank and the bridge as though they were already crawling with Germans. The girls served as "runners," bringing us terse little notes from Bertrand asking about "ammunition levels," for example, or about "troop morale." They also replenished our bread and chocolate rations and took back our "reports." They had to move without being seen, without making the least noise, under pain of being "spotted by the enemy and giving our position away." That was the rule of the game, rather childish in appearance, but we respected it completely, I would say almost religiously, and I remember very well why. I recall the keenness of the emotion that swept over me, never since recreated despite desperate attempts, as I listened for the imperceptible whispering of leaves or snapping of twigs that announced the only real news that matters to a boy—that someone was stealing toward me, toward me alone, and that that someone was a woman. That was the word running through my mind as I sat frozen inside my leafy shelter, straining my ears, heart beating. I repeated it to myself silently and voluptuously: a woman . . . Nothing seemed to me as beautiful as that word, pregnant with a mysterious, inexplicable, oppressive pleasure. Then I would sense a presence nearby, a light panting, and only then would I ask myself the question: Maïté? Zazanne? But already a little of that pleasure was slipping away because it was on the verge of taking solid shape.

Maïté came only two or three times the first few days. The shelter was so narrow that two could not squeeze inside without coming into physical contact, an arm, a leg, a shoulder. I felt her hair brush against me. She didn't bother to observe our ritual, simply handing me Bertrand's notes as though completely indifferent to them, forgetting any "verbal messages," looking at me with a detached benevolence as I sat there huddled up but longing to take her in my arms. Once she said to me, "We look pretty stupid sitting here like this." Another time: "I know you.

You're afraid of girls, of war, of life . . ." I think it was her idea to combine the "runner's" role with that of a Mata Hari. She needed something to make the time go by, to keep her audience satisfied until the hour of truth came. On her last "mission" she planted a hasty kiss on the corner of my mouth and went off to rejoin Bertrand whose "firebase" was up ahead, near the beach by the water's edge, among the last thickets on the island. After that she didn't leave Bertrand again.

I still had Zazanne. Her bargain-basement lavender water gave her away from thirty feet as she crawled forward with what she hoped was Indian stealth. I would softly call out, "Maïté . . . ," trying to hurt her feelings a bit, to show my preference, out of spite.

But she was a nice girl; she would wriggle her soft form into the green shelter and announce "It's only me . . ." with the foolish submissive smile she always wore.

She too was untroubled by ceremony, but for different reasons than Maïté. Its only meaning for her was as a way of getting close to the boys. She would recite Bertrand's "verbal message" as though mouthing Chinese words and then revert to her role of the previous summer, the only opening she knew, adding almost at once in her moist, throaty voice, "I'm frightened, you know . . ." If I was slow to understand, she would persist: "I'm frightened. Make me feel better . . ." Making Zazanne feel better was not a light undertaking. A year had gone by since the La Jouvenière omnibus, and even though I had remained innocent she had made huge strides and was no longer satisfied with sketchy embraces, chance meetings of hands. She overflowed from every orifice. And there at least, I did console Zazanne! I caressed her from head to toe. Stretched out on her carpet of leaves, she had a cunning way of guiding my hand with slow and slight movements of this or that part of her body that was much to my liking. But she kept her eyes closed. She knew me too. I couldn't have moved my little finger with her watching me. If she felt me hesitate she would whisper, "This is war, idiot. Take advantage of it." I don't know whether or not the others "took advantage." Bertrand, definitely not. Pierrot? "How could you think it?" Zazanne protested. "I know him too well. A school friend." Zigomar? Probably. But it seems that it was me she "loved." And why?

"Because Maïté isn't for you and that makes you unhappy. And since Bertrand isn't for me . . ."

There you are. That's how fate operates. That's how the heart's decisions are made, more often than not. Good old Zazanne. The women who succeeded her all through the rest of my life — because someone else "wasn't for me" — probably weren't her equal.

On the last day the girls acted as "runners," when I heard her approach along the footpath, instead of purposely using the wrong name, I called out, "Zazanne?"

"It's me. Were you really waiting for me?"

"Really," I answered stupidly.

It was true. That last day I kept my eyes closed too. And my hands, none too sure where they were going, finally found that triangle of flesh beneath its thick fleece. I found my fingers wet and heard for the first time that labored breathing that marked a frontier of whose other side I was completely ignorant. We didn't cross it. At that age, in those days, it was already pretty good to have gone that far. A memory of war . . .

The war we encountered again each evening when we jumped on our bicycles and dispersed, to meet again next day and exchange news of what we had seen and heard. Pierrot at home, on his farm, where his peasant father was methodically burying everything eatable and drinkable in caches. Zazanne at home with her butcher papa, in the village. He had lived through Verdun, in the old days. Behind his lowered iron storefront shutters he launched into a monumental drunk from which he emerged only a week later, when Marshal Pétain issued his first appeal ("certain of the support of the nation's veterans") to the French . . . but that is another story. Zigomar at La Cornetterie where he consoled his mother, already seeing herself as the widow of the intrepid Lieutenant-Colonel Alphonse Durand of the Pharmaceutical Corps. Finally Maïté, at La Guichardière, where in an orgy of spluttering motorbikes policemen had converged, haggard and exhausted, to inquire after "Mr. Saint-Léger Léger."

"Right out of a bedroom farce, a name like that!" Aunt Octavie declared, right under the poor men's noses. "This is all most assuredly quite grotesque . . ."

Maïté, not wanting to miss a thing, herself led the policemen to the living room. Gloriously draped in the folds of a plum-colored dressing gown, the poet-diplomat had coiled his long body, like a big cat, on the cushions of the largest wing chair. He sipped at a glass of champagne with a greedy darting little tongue. The room was fragrant. A perfumed, youthful flock of aides, frantically wringing white hands, squabbled like lapdogs over the house's only telephone, Petit-Bossay Number 2, by means of which Mr. Alexis Saint-Léger Léger (following Aunt Octavie's lead, we all now called him Léger-Léger), secretary general of the Ministry of Foreign Affairs, whose titular superior was the prime minister in person, Paul Reynaud, strove to arrange emergency lodgings in the region's châteaux and stately homes for the embassies and foreign legations hourly expected in the baggage-train of a government preparing to flee Paris. The policemen scraped the floor with their hobnailed boots. They wanted orders, assignments, addresses, anything that might conceivably resemble organization. They were offered champagne, taken from the bottomless trunks of the Foreign Ministry limousines, which had many other resources to offer. An attaché with fiery eyes was fastidiously spreading foie gras, apologizing to the plum dressing gown for having to do it on such coarse country bread. Gray with dust, goggling, the policemen gulped their glasses down.

"Haiti," one of them kept stubbornly repeating, clinging to the name like a shipwrecked sailor to his buoy. "The Haitian legation. Will somebody tell me where I can find the Haitian legation? I am responsible for the safety of the Haitian legation. Where can I find the Haitian legation?"

"Come, Gérard, jump to it!" said Léger-Léger. "The gentleman is looking for the Haitian legation. Is there anything more important in the world, at this moment, than the Haitian legation?"

Little Gérard giggled, feverishly consulting his files and then the ministry's single Michelin road map, which lay open, crisscrossed with red and blue pencil lines, on a card table. The telephone rang incessantly in a cracked tremolo that grated on the nerves. Two times out of three it was a wrong number. "No! This is not the Ministry of Pensions!" Little Gérard yelped. The bungled connections waltzed around the whole of Touraine from château to château and from ministry to ministry, to the whirring accompaniment of the operators' cranks as they fought a rearguard action against mounting panic.

Whenever we walked Maïté home we took up our observation posts at the bottom of La Guichardière overlooking the main road.

"Let's see what they're up to," said Bertrand.

Nothing "they" did surprised us anymore. The demented procession had been filing past for three days now. The stream had widened and been enriched — if that is the word — by the unlikeliest additions, such as the truck from the Paris Wax Museum from which a host of wax dummies came flying: Maurice Chevalier, Pétain, Lebrun, Tino Rossi, Daladier . . . Our glorious national heritage, rudely jettisoned to make room for leaderless weaponless soldiers who leaped aboard the truck amid the invective of a wretched covey of women trudging along with sick-looking children in their arms. I also remember a full-scale funeral procession, led by a hearse with the deceased's coffin bearing his initial, F, on a silver crest and followed by a small van bristling with funeral wreaths, like a float in a floral parade, and a small black bus packed with some thirty people doubly lugubrious in full mourning, gloves, veils, cravats, the lot, exchanging fearful looks as the mad flow of refugees bore them along, willy-nilly and only God knew where. Torn away from the

family vault they had been seeking, rerouted by fate, the unfortunate F and his loved ones, swallowed up by the current, finding
cemeteries closed and grave diggers gone in every village they
passed, were doomed to wander forever with "them" in expia-
tion of their sins.

We were crouched, hidden, behind the balustrade. Bertrand
raised himself to a standing position. I thought it was out of
respect for the dead. Maïté and I followed suit. From below, then,
we could be seen, should anyone chance to raise his head. And
this is exactly what one of the black-clad mourners did, his
sorrowing mask abruptly turned to fury. The bus windows were
lowered. "Little bastards!" I heard. Thirty faces tilted toward us.
The women had raised their veils, features contorted with rage.
The whole bus seemed to have gone mad. A wrinkled old
woman shook her fist at us. Heads and arms gesticulated at us
from the windows like marionettes. But what had we done to
them? Looking at Bertrand, I understood. Flapping his hands
like wings, his thumbs to his temples, fingers spread, leering
abominably, mouth twisted, he was leaping up and down and
sticking out his tongue at the unfortunates below. He winked at
me, but there was no humor in his eye.

"Do as I do, we'll show them."

Show them what? I had no idea. As usual, I went along, to
please him, to be liked. My most horrible repertoire of faces, the
ones I usually made in the bathroom mirror to tell myself
frankly what I thought of myself. They were ugly. I was very
proud of them. Below, the invective rose to a frenzy.

"Filthy little swine, may God strike you down!"

Maïté contemplated the scene with a distant gaze, her eyes dry,
even when one of the women, a very young girl who must have
been our age, collapsed into tears, hiding her face in her hands.

"Dirty little whore!" a voice cried out, to which Maïté
responded with a deep, formal boarding-school curtsy as the bus
continued on its snail's path, until all we could see was its rear
and a wreath expressing "Eternal Regrets."

Behind it, their noses in the faded mauve chrysanthemums as
if they were following F's funeral, pushed along by the mob

behind them, limped tattered soldiers, the bravest of them still trailing their rifles, then others following them, and then still others as far as the eye could see along the main road. Bertrand regained his composure as quickly as he had lost it. Thinking back to it now, I recognize it for what it was: deliberate theater. He shrugged and turned his back on them.

"F like France," he said. "I'd like to stuff their eternal regrets up their backsides . . . Let's go. We've seen enough."

Between "them" and him there was nothing left in common. Now he had to prove it.

At La Celle, as at La Guichardière, policemen stood guard at the gates in case some tributary from the exodus turned off in our direction with its cargo of broken beings. Flat Top had ordered it: "This is a government ministry, not a first aid station. We must be able to work undisturbed. War means sacrifice. We must steel our hearts against pity. That is the price of—" But at the last moment he quailed and omitted the word "victory." Flat Top was bursting with health; reluctantly admitted to Aunt Melly's table, he was eating as much as four men. Ever alert to the need for "sacrifice" in time of war, he had finally succeeded in expelling the air force mechanics and their anti-aircraft gun.

"Quite right," one of the gunners mockingly approved. "What if we brought a German plane down right on top of your goddamn ministry? Can you see yourself having to fight a war?"

Bertrand and I were there when they packed up. We knew their commander slightly, a reserve officer cadet about twenty years old, smooth-cheeked, with a baby face and big melancholy eyes, which have stayed in my memory although I have forgotten his name. He was Frantz von Pikkendorff's replica, same size, same age, same aristocratic air. The difference was in their eyes. In the cadet's gaze, I remember vividly, were the broken dreams of youth. Every evening during those three days, returning from La Guichardière or from Blue Island, we chatted with him under the cluster of trees where his tent was pitched. He told us the story of his war, which, with wry humor, he called "The Ballad of the Lost Howitzer." He claimed he had heard the Stukas laughing at him. He had seen combat only once, in early June behind the Somme, his gun firing over open sights at a tank formation, advancing as calmly as if on parade, on the other side of the river. Surprise was on his side. He "fried" one of them. A general had raced by on the passenger seat of a motorbike, shouting into the wind: "Well done, my boy, well done, but you're all I have left! We're moving back! Evacuating!" Later, despite his protests, he was packed off to Touraine to provide air safety for an airbase that existed only on paper, with no planes, just a handful of mechanics who killed time playing *belote* — but still put the fear of God into Flat Top. On the barrel of his gun he had painted a single tank silhouette, followed by three periods . . . He was being sent away. He left.

"And if you were to *fry* another one?" asked Bertrand as the cadet packed up and climbed into the cabin of his truck.

"Oh yes? Where? And all on my own? Like a grown-up? A private war? For whom? And why?"

"On my island!" Bertrand said gallantly. "For me!"

It could have been absurd. It wasn't. When Bertrand put on the voice of Bertrand Carré, son of Squadron Leader Carré, stationed on the Chinese frontier, his words fell from so high above it didn't occur to you to laugh. Then he spoke about Blue Island; he drew a map of it on the ground. The cadet listened gravely, the way you would listen to a child if you had any recollection of your own childhood. The cadet's was not so very far away. There was a brief moment of harmony. Then the cadet gathered up his marbles (I know no other expression that conveys it so well), and Bertrand found himself alone. The other had gone back to his lost universe, his ruined dreams, reality. He saluted us briefly and perhaps a little apologetically from the window of his truck as he pulled away. He seemed sad and resigned.

"A pity to be so old already, at his age," Bertrand commented simply. "Let's not mention him again. I misjudged him. He was no different from the others."

The underbrush was littered with cigarette butts, soiled paper, bottles, and empty bully-beef cans. The aircraft engine hung from its tripod. They hadn't even taken it with them . . . We headed back to the house.

"Ah, here are the boys!" Aunt Melly called out brightly.

"The boys" were Bertrand and me. Although I had scrubbed myself when I got in, I was sure I reeked of Zazanne's cheap lavender water. I already felt that bleak sense of something wrong, the feeling that has always swept over me after leaving the arms of women who were mere caricatures of my dream. Thank God Bertrand overshadowed me, striding forward draped in innocence.

As always in late afternoon when it was hot, as was the case that June 10, a little predinner buffet of lemonade and chilled Vouvray had been set up under the trees. My mother thought it "wasn't decent, with all those poor people on the road," but Aunt Melly insisted. A final memory of peacetime, just one more minute, Mr. Executioner . . . The shadows of the older cousins and younger uncles lingered like a faraway echo around Aunt Melly Lavallée, thoughtful, solitary, stretched out on her lawn chair. There was no news of anyone except Uncle Sébastien, who, from The Three Ducks in Loches, was in a stupor of disbelief as he watched what he still called "the front" (as in 1914) inexplicably approach. So our families were closing ranks. The Majorels, who had no ministry to entertain and whose radio was broken down, had come as neighbors to get the news. Flat Top was also there, for his "office" shut down at six—with the exception of the telephone operator—a model of administrative continuity in wartime. My cousin Louis Lavallée had invited one of Flat Top's typists, who addressed him as "Mr. Louis" as if she were a maidservant and who lifted her little finger so high in raising glass to lips that it was hard to believe it was attached to her hand. An unexpected opportunity for him. He was eighteen years old, he was very ugly, and Aunt Melly did not like him. She kept shooting him horrified looks, but when all was said and done his dinner guest did add decorative value.

My mother was fretting. "We don't ever see you anymore, what do you do all day?" she asked. "Is it proper for you to be disappearing for hours on end at a time like this? What if something were to happen to you? I hope at least you're behaving reasonably?"

Cousin Louis sniggered. Aunt Melly lost her thoughtful look. She rose elegantly from her lawn chair.

"We can trust Bertrand," she said with a barely discernible smile that belied the assurance of her remark.

I never really noticed that smile until forty-seven years later, when my memory played the scene back to me. That day Aunt Melly was wearing a light summer dress, red with small white polka dots, fairly low cut, bare arms, very young. Recklessly, Flat Top launched himself into a torrent of high-flown compliments.

"That dress! A glimmer of joy amid the gloom —"

"Don't! Just don't!" Bertrand spat through clenched teeth as he walked by.

The man was speechless. Later I heard him complaining to the Majorels and my mother that he had been "stabbed in the back." He wasn't talking about Bertrand but about Italy's declaration of war, just announced on the radio. His anger was volcanic. He was taking it out on Mussolini for being snubbed a second time by Bertrand. The rest of the news was no better. Uncle Armand was taking comfort in the wine.

"It's beyond me!" he said. "It's beyond me! What is Weygand doing?"

Out of curiosity, I have checked. On June 10 General Weygand was laying down the fateful word *armistice* on the green baize tabletop at the last war cabinet meeting to be held in Paris, a word that was on the tip of everyone's tongue but that no one had dared to utter . . .

"Come, you must be thirsty," said my aunt.

She slipped one arm under Bertrand's and the other under mine, her hands closed around our wrists, and I believe her fingers were toying distractedly with Bertrand's bracelet. I was red as a beet as she led us to the buffet. I remember her cool skin against mine, the gentle possessiveness of her hand, but, as I have said, it was just habit with her, she meant nothing. She left us there with no further ceremony, glasses of lemonade in our hands. From the castle, the aide manning the phone was running toward us. My father had just called from his ministry, and what had just happened was no longer important.

The government was leaving Paris.

That night we stayed up late. The Majorels and Bertrand remained for dinner. Following this new twist of fate no one wanted to be alone. The radio was never switched off. At around eight that night an appeal by Prime Minister Paul Reynaud — the man my father called the "pathetic dwarf" — was announced. And he certainly had a dwarf's voice, I remember: "In the course of her long and glorious history, France has survived harsher trials than this." ("I wonder which ones . . . ," my aunt commented.) "It is at such times that she has surprised the whole world. France can never die . . ." Then came the inevitable *"Marseillaise"*: it was played a dozen times a day, its martial accents a heroic counterpoint to the long litany of our disasters.

"You know what they should play instead?" Bertrand whispered into my ear. *" 'Madelon'!"*

At eleven p.m. a brief communiqué was issued: "The government has been obliged to leave the capital for urgent military reasons. The prime minister has gone to take his place with the Army."

Around a quarter of an hour later, my father called from Chartres police headquarters. The government convoy had left Paris three hours before the communiqué was issued, lights out, like criminals, and using back roads for fear of being swallowed in the torrent of refugees the communiqué would inevitably set in motion. It had nevertheless taken them all that time to travel the sixty miles from Paris to Chartres. Just before Paris, Normandy had burst open like an abscess, its refugees overtaking Picardy, which was already treading on Flanders's heels. At

Chartres the convoy had been heaped with insults. The town was packed with fugitives, and its poorer neighborhoods had been looted, with grocery stores pillaged and empty houses broken into and robbed. The riot police had been forced to fire into the air to extricate Miss Lily Palma, Chabannais's *éminence rose,* who had been spotted amid her hatboxes and travel cases in one of the official cars. They were preparing to get back on the road, my father told us, but it was difficult to say when they would arrive. As for the pathetic dwarf, he had not taken his place with the army at all. That was theatrical bragging, lies. On the contrary, he and his little convoy had turned their backs on the army the better to make their getaway . . . Then my father was cut off.

Reporting the conversation to us, my mother's only comment was, "What a pity!"

"Poor France," added Uncle Armand. "And who is this Lily Palma?"

Bertrand and I, lying low in a corner, didn't miss a word of what was said. My mother mentioned the *cabinet rose,* or "pink cabinet," an expression I had already heard at home without any clear idea of what it meant. I think it was my father who coined it, and it was quickly picked up by chroniclers of that sad period. It referred to the mistresses of all those gentlemen, Reynaud, Mandel, Daladier, Chabannais, government ministers, leading statesmen, senators, deputies. At this mass exodus of the nation's rulers it was as if a henhouse had opened: they streamed out from all the elegant neighborhoods of Paris, commandeering the best seats, taking over the biggest cars . . .

"Kept women," my mother said, "tarts, actresses, singers, dancers, divorcées, even a marquise and a countess — they get in everywhere with their dogs; they have opinions on everything; they scheme; they make demands . . ."

"Poor France," Uncle Armand said again.

And it is true that when Paul Reynaud resigned a few days later, in Bordeaux on June 16, leaving Pétain a free hand, it was to the Countess de Portes, the dwarf's mistress, that France owed undying thanks . . . But that too is another story, told to me much later by my father . . .

"Poor France!" Bertrand added, parodying Uncle Armand in a whisper. "Do you feel any pity?" he asked me.

I naturally tailored my answer to his scathing tone. It is true that we had all taken a heavy pounding these last three days, seeing much and hearing much, and that we were only fourteen. A government that runs away in a flurry of skirts, a head of state who lies and then bolts like a rabbit, a people that loots as it flees – these were the embodiment of chaos. Yet I believe it would all have slid past me, leaving me indifferent, mildly surprised (as the five subsequent years of war would slide past me), had I been alone. Instead of which, I put on a look of disgust.

"Bunch of fairies . . . ," I said.

Bertrand's blue eyes caught me like a blow to the chin.

"Wait till it's our turn to show them! Are you ready? We won't have to wait long for these Germans."

We would have to wait another ten days. I know it, because I counted every one of them. Ten days of bragging as terror ate my insides. But, God! those ten days went quickly!

WE GOT UP early on June 11. The weather was delectable. The sun shone, and the trees were more murmurous with birdsong than I have ever heard them since — as if the entire feathered species had assembled for an exodus to parallel our own. "I don't want you wandering off today," my mother had told us. "I want you to be here when your father arrives and not to leave the house again." So Bertrand came to join me, soon followed by Maïté, Zazanne, and our whole group. Like a line of birds on a telephone wire, we sat swinging our legs on the elegant stone balustrade of the front stairway, eaten up with curiosity, dying to see what Miss Lily Palma, of the influential pink cabinet, might look like. According to my mother, she (through the intermediary of Chabannais, "that radical rustic intriguer to whom we've entrusted our industrial effort," as my father had recently described him) was responsible for all of France's misfortunes.

"Do you want me to throw her out?" Aunt Melly asked. "After all, this is my home."

"What's the use?" my mother replied. "All this will be over soon. Gustave says so." (Gustave was my father.)

They were waiting there as well, their eyes red after a sleepless night. Flat Top was pacing the gravel driveway looking careworn and casting worried looks over his shoulder at the six boys and girls sitting in a row as if to watch a show. Or like a row of magistrates in court . . .

"What are you kids doing there?" he said aggressively. "Isn't the garden big enough for you?"

He seemed truly ill at ease. Second-rater though he undoubtedly was, I imagine that a little shame was beginning to stir in his subconscious at the prospect of welcoming a member of the French government in headlong flight, his mistress in tow, before the pitiless gaze of a group of fourteen-year-olds, whose ringleader's arrogant eyes reflected contempt for the world in general and for Flat Top in particular.

"And you!" he said, pointing at Bertrand and using the familiar *tu*. "You think you're tough?"

Bertrand's answer cut like a knife, "Not *tu*. *Vous*. No one speaks to me so informally without permission, and you don't have my permission."

"Nor mine!" I said, close on Bertrand's heels, but a look from my mother made me lower my head. Such parts weren't for me, and everyone knew it.

Thinking back to this scene, I still wonder why Flat Top didn't just box Bertrand's ears. It was because he *couldn't*. Bertrand towered so far above him that he didn't even make a move in that direction. Instead, unable to wriggle out unaided, he turned distraught eyes toward the lady of the house and implored her help.

"You . . . You allow this, madam?" he managed to stammer out. "When the minister hears . . ."

A faint smile on her lips, Aunt Melly remained silent. She looked at Bertrand, and her eyelids narrowed slightly as though she were encouraging him.

"The minister! Exactly!" Bertrand called out, like Ruy Blas addressing the queen of Spain's ministers, a scene he knew by heart and often recited to me. "I have never yet seen a minister. I don't want to miss this one. I want to know whether he's worthy of this mess, if he's as disastrous as this disaster. I'll know the second I lay eyes on him. I am never wrong."

"And then?" Flat Top asked mechanically. He was defeated.

"Then . . ."

Bertrand jumped from his perch and landed in front of the fat man, who stepped back instinctively. I sensed the approach of the speech.

"Then," the boy went on, "then I, Bertrand Carré, only son of Squadron Leader Carré of the foreign legion, stationed on the Chinese border, and of Laïcha, Oriental dancer, who has given her son to France, fourteen years old, crack shot, knife thrower, acute nocturnal perspicacity, volunteer for any dirty work, defender of the Blue Island marches, and in love, that goes without saying — I will erase everything and begin again on my own behalf . . ."

And gesturing with his arm, the bracelet glittering in the sunlight, he addressed a small salute to my aunt and another to Maïté, like a knight cantering into the lists. The two women reacted in opposite ways to that surreal act of homage. Aunt Melly smiled, delighted. Without question, this unpredictable young fighting-cock amused her. What had he come up with now? She clapped with the tips of her fingers to show her pleasure. But it is Maïté's face I remember best. A serious, almost tragic face, turned toward Bertrand as though drawn by a magnet, her gray eyes never leaving his lips as though she sought there, for herself alone, the true meaning of his words . . . The rest of his remarks were lost in the roar of engines at the main gateway. Two dust-coated motorcycle policemen appeared at the bend in the drive, followed by a line of official limousines. I counted six of them. The Honorable Minister for Industry had arrived at Château de La Celle, temporary seat of his power.

Without another word, Bertrand returned to his place on the balustrade and waited, elbows on his knees, chin in his hands. It was then that Flat Top, making for the limousines as they halted one after another, suddenly turned back and came over to where we were sitting. Addressing Bertrand in a submissive, almost humble voice, he said what was at the very least strange: "Please, try to understand . . ." Was he afraid Bertrand would make a scene? I don't think so. This time Flat Top wasn't alone: my father, at that very moment getting out of the first car, would never have tolerated such behavior from Bertrand, since the "radical schemer" was after all his superior. It is only today, of course, that I realize the unusual nature of Flat Top's plea (to which there was no reply). Somewhere inside his truckling soul, embers of courage must have remained alight, some capacity for judgment of character must have endured — and therefore some sort of esteem for Bertrand, prompting Flat Top to deliver a disguised warning. Out of friendship? Why not? When you judge someone too harshly you are almost always mistaken. But I have pursued my thoughts on the subject no further, since from this moment on Flat Top ceases to play even the smallest part in the story. Bertrand saluted his departure. As soon as the fat man had turned his back on us, resuming his flunky's bearing and stationing himself beside the opening door of the ministerial limousine, Bertrand whistled between his teeth and pointed at the car.

"Understand what? *That?*"

A dog was the first to emerge from the car, a grotesque white poodle, body shorn and neck and paws left thick and curly, a tuft of hair on the top of the skull and another at the tip of the tail. The two motorcyclists had dismounted and were watching it with undisguised hatred. Frozen in rider's postures, legs bowed as though their joints were stuck, they had bloodshot eyes and reeked of sweat and leather at twenty paces. A night on the road with whistles clenched between their teeth, to escort "that," to open a passage through terror-stricken crowds for "that" — how many had they bowled over, even run over, in the darkness? — a night of being insulted as they shoved their way through the poor bastards crowding around the public fountains in order to fill a canteen with water for "that"! Then the rest of "that" began to emerge. The leash grew taut, and a hand appeared, covered with rings, with mauve nails, and then, like an extension of that hand, a creature hung with pearls from chin to belly, a rounded figure set off by a close-fitting dress, mauve to match the nails and the makeup on the wide mouth with the fleshy lips. This at once came to life as Miss Lily Palma, who took in the aristocratic old house with not a groom or huntsman in sight, and called out in the accents of a Folies-Bergère singer (which is exactly what she was), "Well, now, this certainly isn't Deauville! Clapart, my angel" — she was addressing the ponderous Flat Top — "Mistouflet and I desperately need a bath."

Mistouflet was the poodle. Not the minister. We could be certain of that because a second later she turned back to the car, in which nothing was stirring, and said, "Riri, we're here." Then, explaining: "He's asleep, the poor darling. He's exhausted. Oh, it's not funny, being a minister when nothing works the way it used to!"

Obviously . . . We heard a light laugh. Aunt Melly had decided to be amused. She walked down the few steps of the stairway. "Clapart-my-angel knows the house," she said in her most snobbish voice. "He will show you to your room . . . Miss."

"Riri, come on!" Miss Lily called again.

"All right! All right! I'm not deaf," a sleepy voice answered.

At long last the honorable Henri Chabannais, minister of industry in Paul Reynaud's war cabinet, emerged from the car: two hundred and fifty pounds of fat and bone draped in an ivory silk suit, double chins, a triple fold at the nape of his neck. The poodle yapped and made ungainly leaps around him.

"Quiet, Mistouflet," the minister said. "You'll be fed soon."

"That he will, the darling," my aunt agreed in sugary tones. "In the kitchen or in your bedroom? Only one bedroom, of course, Miss Palma's."

Chabannais was suddenly fully awake. We didn't miss a word.

"Of course, of course," he muttered, screwing up small, furious eyes.

Pierrot counted the suitcases, flabbergasted, mouth agape: luggage from Vuitton, from Hermès, portmanteaux, hat boxes, vanity cases, holdalls. The chauffeurs of the other limousines had rallied round; even Flat Top labored under the weight of two heavy overnight bags.

"Ah," said the minister, remembering something important, "can you provide quarters for my cook?"

"Your cook? Of course," my aunt answered, "but we haven't much room. In the kitchen or in your bedroom?"

Chabannais's gleaming skull had broken out in red splotches. His double chin trembled. He hesitated, then sighed.

"Please, Madam, I'm tired."

"Of course," said my aunt, but no pity touched her voice.

The minister opened his mouth to reply, then closed it again. He walked heavily toward the stairway, in silence, followed by Miss Palma, her posterior swaying spectacularly at every step, and his baggage carriers. He was not very tall, but massive, thickset, all meat and stomach, his forehead low, with fat hairy hands and an indecent square signet ring on his little finger, the image of the sly political hack from a rural district who had slipped effortlessly from the humbuggery of the market square to the chicanery of parliamentary life. My father walked close to him but didn't carry a bag. Although taller by a head than the minister, he did not once stoop as they spoke. I think my father was a kind of honorary minister, addressed by the simple title of *directeur,* but in reality representing the interests of French corporate and industrial management.

As he walked past us, Chabannais pointed with his chin and asked stiffly, "Who's that kid there?"

"That's my son," my father answered, taken aback, directing a little wink at me that meant: don't worry.

"No. Not him. The other one."

This time his tone was openly hostile. I turned toward Bertrand and saw that his eyes were fixed on those of the minister. How long had he been holding that stubborn stare? Ever since the minister had appeared, no doubt, and the fellow had noticed it. Had Bertrand not maintained that he would judge him with a single glance and that he was never wrong? The verdict had been reached, and it was uncompromising. It stared as plain as writing from Bertrand's blue eyes, from which all warmth had drained, giving place to regal scorn.

"A friend of my son's," replied my father, and said no more.

The minister stopped for a moment. I noticed that all of us— my aunt, my mother, the two policemen, the drivers, the secretaries, even Flat Top—had our eyes fixed on Bertrand's: none of us, given the circumstances, could fail to interpret what Bertrand had concluded—that this man and everything he represented deserved his own weight in contempt. The red blotches reappeared on Chabannais's forehead. But when he met Bertrand's eyes it was he who lowered his gaze. I thought he was going to speak. He merely shrugged and disappeared into the house.

Miss Lily had noticed nothing.

"Those boys are so sweet," she simpered.

She gave off a powerful fragrance, a rich blend of her own redhead odor, for she had clearly perspired abundantly during the journey, and of some costly perfume from the rue Saint-Honoré. Scarlet to the roots of his hair, mouth open, unable to breathe, Pierrot contemplated the most beautiful creature on earth. As for me, for once in my life I stared openly and without embarrassment. Miss Palma's vulgarity, the unmistakable nature of the sole employment for which she was designed, brought her down to my level without my even realizing it. The realization was to come to me later, with others . . . Split at the knee and low-cut, the mauve dress revealed the thighs of a Flanders mare and a ponderously swelling belly. You would have said Zazanne fifteen years older. In fact, Zazanne whispered, "I'm jealous," like a pulp-romance heroine.

"And what's this young man's name?" asked Miss Lily Palma. She naturally referred to Bertrand.

The "young man" shot her the same icy look he had given the minister, but fleetingly, as though she were not even worth his scorn, and with just a hint of the salacious thrown in, bringing Miss Palma still lower and compounding her confusion with a touch of self-satisfaction that went straight to her loins and provoked a fatuous smile.

"Please. Not that," Bertrand said tonelessly.

Then he ostentatiously turned his back on her, leaped from the balustrade, and took off for the corner of the garden where our bicycles were lying. We followed him. The expression on his face told us this was a bad day. When he looked like that—and when he did we rarely had any idea why—we stuck to him like flies, surrounded him as if we were his courtiers, for if we didn't he would drop us flat, and without him we were nothing anymore, wandering in boredom. This time we were unlucky.

"Get out of here!" he ordered. "I don't want to see you anymore."

Maïté stood next to him. His anger was never directed at her. The two of them mounted their bicycles.

"You," he said to me. "Follow us."

"I'm not allowed to leave the house."

"We're not going far. The rest of you can go to hell!"

We pedaled for a few minutes. A forest road, then a footpath. Bertrand leaned his bicycle against a tree and slipped through the ferns with Maïté at his heels, beckoning me to follow. Twenty paces farther on we broke into a small enclosed clearing, ringed by trees whose foliage framed a vault supported by trunks straight and tall as cathedral columns. Clearly they had both been there before. Bertrand nodded to Maïté. Between two slender trees was a broad stump sawn off close to the ground, making a kind of pedestal. Maïté approached it, removed her clothing nimbly, her pants, her bra, her sandals, folded them, hid them in the grass so that there was nothing to indicate that she had ever worn clothes in that clearing, then leaped onto the stump and waited, arms at her sides, thighs together, eyes closed. When that scene comes back to me, what I notice first, disconcertingly, is the extraordinary absence of affectation in the two participants.

"Now look at Maïté," Bertrand told me. "Look but don't move. Imagine you're in a temple. Take your time. Direct your eyes, lead your soul where you will, over every detail of her body. You must not miss anything about her. Everything there is perfect."

Rearing up on bare feet, raising her arms high above her head, hands outstretched as though she were offering herself to the firmament, she began to turn round and round. Bertrand had used the word *look*. He could easily have chosen another: contemplate, admire, worship, revere . . . I have already described Maïté in that month of June, 1940. I will add nothing to that description, except to say that in that moment something divine seemed to emanate from her, a wordless hymn, a heavenly fragrance only my emotions perceived.

"A good reason for dying, don't you agree?" Bertrand asked me.

I stared at him, stunned. Yes, he was talking about Maïté.

"Who — who wants to die?"

"We're at war, aren't we? But nothing and no one is worthy of it. Everything is so ugly . . ."

Maïté was facing us once again. She looked at me. She smiled at me. She had come back to earth.

"Certainly," I ventured. "But surely Maïté is a reason for living?"

"You'll never understand anything," he said to me.

Then, as Maïté dressed again, he changed his tone. "Well, have you erased her now? Did our shock treatment work?"

I was understanding less and less.

"Erased who? Erased what?"

"The one who made such an impression on you, Miss Lily. I watched you. You couldn't get enough of her. You would have carried her bags and stroked her poodle just for a walk in the woods with her. Listen to me. I'm your friend. What I am about to say will be useful to you later on. I don't want to let you make a mistake. There are two kinds of women. The rare kind, who do a man honor. And then the others, the ordinary ones, who do him dishonor. I wanted you to understand the difference once and for all. Never forget it."

I don't know whether he was conscious of the cruelty of the picture he painted, even if it wasn't truly intended for me. In fact, that is precisely why I had wanted to forget this moment . . .

I went home alone.

By noon Chabannais was back on the road. Only one car this time, two policemen but no dog, his ministerial portfolio, and a hamper packed with provisions. He also stole my father away from us. Their destination: Château de Cangé, in a suburb of Tours forty-five miles north, where the cabinet was to meet. He left us Miss Lily and his swarms of aides and flunkies who flowed into every nook of the house in search of a corner they might call their own. We had had to put out extra beds, lay mattresses on the floors, get old settees down from the attic, set up screens to protect the typists' virtue. Bathrooms were besieged. Miss Lily locked herself in hers to avoid having to share it. Lines formed outside toilet doors. Voices rose in protest: "What's taking so long in there?" The gentlemen went out to pee in the garden, watering my aunt's rosebushes. Some commandeered deck chairs and snoozed in the sun or else picnicked on the lawn while two under-strength teams put on a display of soccer. I wonder where they found the ball: soccer had never been popular with the aunts. Making do with whatever came to hand, they rebuilt a tolerable life-style for themselves. No longer serving any purpose, they had adapted to their uselessness with disconcerting swiftness. Even the telephone had stopped ringing. The switchboard operator sat in front of her one mute phone filing her nails. Indeed, who would have guessed that amid the general collapse there was still such a thing as the ministry of industry, or any other kind of ministry for that matter? Flat Top alone maintained the fiction, toiling away in his improvised office in one of the drawing rooms, poring over ten-day-old mail that had only the most surreal relevance to the nation's situation.

But at five the telephone jingled. It was my father, calling from Cangé. He wanted to speak to my aunt. The cabinet meeting was over. He was getting ready to get under way with Chabannais, by a roundabout route that would avoid the refugee flood. The minister wanted to give a dinner that very evening. My aunt thought she would burst.

"A dinner! At this hour! At my house! For whom? How? With what? There's just no sense in it!"

But it seemed that there was. The chauffeur had commandeered supplies from the stores and cellars of Cangé police headquarters. The car's trunk was overflowing. My aunt was not to worry, the minister's cook would improvise.

"War is war, is that it?" my aunt said nastily. "It's out of the question!"

"Melly, the minister is only trying to do what's expected of someone in his position. He just asked me to tell you that," my father added hastily.

It was the inappropriateness of this argument that gave it victory. By a knockout. My aunt gave in. Twenty places were laid for nine o'clock. Chabannais had invited Alexis Léger, Uncle and Aunt de Réfort, his hosts, another minister billeted at Grand-Pressigny, a few diplomats from various houses scattered through the area, a monsignor from the papal nunciature . . .

"And who will preside over it all?" my aunt asked suspiciously.

"The minister, of course."

"I know, but opposite him?"

"You, Melly, naturally."

"I see . . . And Miss Palma?"

"On the minister's right, Melly. He always strokes her thigh under the tablecloth. He likes that."

That conversation is part of family folklore. Which is how I came to hear of it . . .

Picture the situation. The country was rapidly decomposing, its head already beginning to smell of putrefaction; its capital had been declared an open city—the radio had just announced the news—and would not be defended; a million prisoners had fallen into enemy hands, whole regiments minus their leaders surrendering to corporals on motorcycles and falling into docile columns; the Germans had crossed the Marne and were preparing to enter Paris; and the roads to the south had become human floods beyond any semblance of control . . . And in the midst of all this, instead of staying together and presenting a united front to the situation, the leaders of France persisted in their absurd

living arrangements, leaving the cabinet meeting singly to return
to quarters scattered over the countryside, from one end of the
Indre-et-Loire region to the other, to bury their fear inside their
temporary châteaux or to go on weaving their webs of lying and
intrigue and even, yes, to throw dinner parties! Georges Mandel
at Château de Sauvigny, Camille Chautemps at the manor house
in Druyne, Daladier near Montbazon, Chabannais at La Celle
. . . "What worried them most," my father told me later, "was
what would become of them personally. How they could make
sure they landed on their feet when the Reynaud government
collapsed, which was only a matter of days. A few were already
calling Bordeaux to make sure of the best accommodations and
to have them furnished and repainted as befitted their rank.
Others, more farsighted, were eyeing overseas ambassadorships,
assignments that would get them out of France. At the time of all
those shadowy dinner parties they thought they could still pull
strings, plot, call on the services of those beholden to them. They
were hopelessly thick-headed . . ."

I too remember that dinner. The château was blacked out,
scarcely any light filtering through its closed shutters. A ministry
footman wearing his chain of office, a flashlight in his hand,
greeted the guests on the stairway. The salon and dining room
had been set apart for them, while the small fry of the ministry
had been relegated to the other rooms where Chabannais had
laid on champagne to soothe their resentment. In the salon, one
lamp out of two had been switched off, for the Honorable
Chabannais, no hero, believed himself personally threatened by
the Luftwaffe. Human shapes emerged from the twilight, features haggard from fatigue, eyes red, for some of them had driven
all night; nevertheless, they would not have dreamed of missing
this posthumous celebration of their political vanities (another of
my father's expressions). I see it now: scenes right out of Fellini, a
much more recent point of comparison than the events themselves, yet one that exactly translates that night's atmosphere.
Léger-Léger and "Little Gérard" made a conspicuous entrance,
in dinner jackets and wing collars.

"The last sigh of an expiring world, Madam, let us make it memorable . . ." said the famous poet and all-powerful secretary general to the minister of foreign affairs as he kissed my aunt's hand, to excuse the studied flamboyance of his attire but without giving a thought to the defeatism of his words.

Little Gérard looked more cheerful, cheeks pink, eyes shining. I heard my mother whisper to herself, "He's wearing makeup! Poor France." At the time I didn't see the connection.

The other "Excellency" present was a certain Chicherais, minister of trade, I believe — a small gray man with a pince-nez perched on a long snuffling nose and little watchful eyes — flanked by a legitimate spouse who sighed with every step she took. Chicherais's overriding worry was the five sealed trucks that had made the trek from Paris with him and were causing him boundless concern. As a precaution he had had lined them up under his windows at Grand-Pressigny. Members of his cabinet were obliged to sleep on the driver's seats.

"I asked Mandel for a police escort, but he wouldn't listen — you'll back me up there, won't you?" he added to Chabannais. "Can you imagine if it fell into the wrong hands . . ."

I assumed it was the gold reserves from the Bank of France I had heard people mention, or treasures from a museum. But all that the trucks contained were the records of the Radical-Socialist party, to which both ministers belonged.

I heard my father whisper to my aunt: "Five trucks heaped with scandal, tons of nauseating documentary evidence, twenty years of mutual backscratching and compromises, enough to dishonor half the country's politicians, they're counting on it all to keep them afloat . . ."

The Majorels were there as well, Aunt Sophie and Uncle Armand, with a tall, sad Negro rolling doleful eyes at their side. Mr. Agathin Perpétue, minister plenipotentiary of Haiti, who had finally reached La Jouvenière after drifting, lost, all night long. Almost merging with the shadows, he wandered from one group to another, listening silently and nodding, utterly overwhelmed. On the news that night: "The final collapse of the French defensive positions" — in other words headlong retreat —

Churchill's visit to Tours and his refusal to commit the Royal Air Force to the struggle, and France's appeal to the United States.

The air was full of opinions like the following: "Everyone is trying to lie to himself, even Mandel who's walking around giving himself airs.* The armistice is a certainty, but Reynaud will be passing the baby—('the baby!')—to Pétain, and since Pétain hates politicians," Chabannais concluded, "we'll find ourselves on the beach . . ."

"What about Roosevelt?"

"Words, words, that's all we ever get from him. France is going under like a drowning man, that's what Reynaud told him, the coward . . ."

And there was no doubt: fear, apocalyptic fear, was in the air. Agathin Perpétue took out a handkerchief and dabbed at a real tear. Bertrand told me that since arriving at La Jouvenière he had monopolized the telephone, dictating long telegrams to the village postmistress. They were addressed to Port-au-Prince, and they urged his government to declare war on Germany and on barbarism lest history record that Haiti, the world's first black republic, had abandoned, friendless and dying, the mother of civilization and the arts, the repository of light, the home of freedom, of human rights, and of Victor Hugo, the beacon of humanist thought, and so forth.

"He's losing his head," Bertrand had added, "but he at least is losing it with grace . . ."

Indeed, Mr. Agathin Perpétue was offering himself to save France just as our little gang had attempted to save Poland the previous summer, in the bus at La Jouvenière. I remember him with feeling. He was one of us.

Fear gives you an appetite. Everyone fed his anguish. The dinner was Rabelaisian. Geese fattened right up to the moment of their demise. Aunt Melly had set up a small table apart from

---

*Interior Minister Georges Mandel was disliked because he was tough and ambitious, and perhaps because he was Jewish. He was one of the best of Reynaud's cabinet. The Nazis and Vichy later implicitly acknowledged this when they arranged for him to be "shot while trying to escape." — TRANS.

the rest where Bertrand and I sat under orders not to be noticed, along with my cousin Louis Lavallée, who took advantage of the situation by leaving, during the second course, to join his wartime love—Flat Top's secretary, the one who called him "Mr. Louis"—in the woods. Good riddance. I think his departure gratified my aunt. Once her son's gloomy countenance had disappeared she could direct her gaze toward Bertrand. The position of our table made this easy, and I believe she had engineered it deliberately to create for herself a visual refuge, an oasis of honesty on which to rest her eyes. She scarcely spoke or touched her food. She was condemned to listen to Chicherais, who had been seated at her right and who held forth exclusively on himself, telling her in his rasping voice how he had stopped Pétain in his tracks at the cabinet session (although in a craven turnabout a month later this intrepid politician would vote full powers to the marshal, but that too is another story . . .). From time to time my aunt's eyes would slip away and travel over her guests' shoulders to meet Bertrand's. He never talked to me about the unspoken complicity between them. I think it served to reinforce his pride and his growing conviction that between himself and "them" there was no longer anything in common.

Them . . . Chabannais had placed the monsignor on his left and thus, as he wittily remarked, found himself seated between two gowns. He plied the prelate with tales of his own virtue. According to him (and with the help of the good wine of Chambertin), there was no better Catholic than he, appearances to the contrary (this last to exculpate him for the Lily whose thigh he was squeezing as he spoke). He had married his son in church. His daughter had been raised in a convent. His wife, for he was not divorced and made a point of mentioning it, regularly attended Mass. All of which, in a secular republic, amounted to more than courage, it was defiance! And had he not conferred the Legion of Honor on two bishops and a friar during his stint at the colonial ministry? That was worth a little something, wasn't it? A carefully concocted report to the secretary of state at the Vatican? The Honorable Henri Chabannais could very well see himself as French ambassador to the Holy See, safe and warm at

the Vatican and keeping an eye on events as his country crumbled
. . . Embarrassed, the prelate said nothing. Then another voice
captured the table's attention as Miss Lily Palma began to over-
flow with the kind of tired backstage anecdotes — petty rivalries,
spiteful acts, gossip, affairs — that are grist to the theatrical mill.
She was distinctly tight, and Little Gérard was spitefully egging
her on, knowing she would soon pass out. After a while hers was
the only voice that could be heard, her quavering soprano
drowning out all other conversation.

"Quiet down a bit, Lily, you're drowning us," the minister
cried, seeing his Vatican post compromised.

"Fair's fair, take your hand off me," she answered, tit for tat.

In the chilly silence that followed, a distressed and diffident
voice, almost apologizing for itself, said, "And while this is
happening . . ."

It was Agathin Perpétue. His black face had turned gray, his
way of going pale. Tears ran down either side of his broad, flat
nose. His mouth was open, but he was unable to finish his
sentence. Emotion had choked him. We never did hear what he
wanted to say. It was easy to imagine. My aunt Sophie, sitting next
to him, laid her hand gently on his. The Haitian had taught us all
a lesson, and everyone lowered his voice.

Other groups formed when the coffee was served. Everyone
was obsessed with the same question: What was to be done?
How were we to get out of this? Not the nation, but themselves.
I must acknowledge that they were true to character. The political
man who no longer lies has reached rock bottom. Alexis Léger
had cornered the consul general of the United States, representing his ambassador, who had remained in Paris to look after his
countrymen's interests. I didn't hear what they said to one
another. A few days later, Saint-Léger Léger embarked discreetly
aboard an American ship destined for the United States, where
he lived out the war, but that again is another story . . .

"Come on," Bertrand said. "Let's go. A bit of fresh air . . ."

Someone was sitting on the front stairway in the darkness
looking out into the night. I remember those wartime nights

when only the stars shone and there was no light to betray the presence of human beings. At times I miss them.

"Ah, boys," said Mr. Perpétue's voice. "Come and sit beside a broken man. This country I have loved, this great country I have admired . . ."

He put his arms around us as if we were his own sons.

"Have you read *Fermina Marquez?*" he went on with no transition.

We confessed our ignorance.

"In that work your fellow countryman, the great writer Valéry Larbaud, describes in stirring terms the fate of young Haitians of good family sent at great sacrifice by their parents to attend provincial schools in France—gloomy, chilly, damp institutions, but less expensive than Parisian ones—to receive the French education we prized more highly than any other. Their exile from their native land lasted years. Left to cope with your climate, many of these lonely, unhappy children developed heart-rending coughs; they spat up blood, lost their appetites and their animal spirits, but they fought on in silence, uncomplainingly, until one day they lay down to die, like poor lost dogs. Dying for France, in a way, for French civilization. Some survived, of course. I am one such survivor. Six years at the Pontarlier School where Toussaint-Louverture* died a prisoner, heartsick, numb with cold. Proof of our unshakable loyalty. But when I returned to my hometown, Jacmel, cut off by mountains and reachable only on muleback, I was given a hero's welcome. I had seen France! Seen Paris! Spoken French in France, and with Frenchmen! In the street, people fought to touch my garments. I was paraded through the shabby salons of the educated families of Jacmel to show off my learning—even to the *Cercle Littéraire*, whose president sent every newly elected member of the French Academy a telegram of congratulation that most often went unanswered. We passionately dissected the few newspapers sent from Paris, three months out of date. I had to recite poetry, *Ruy*

---

*Born François-Dominique Toussaint, Haitian general and liberator, circa 1743–1803. —TRANS.

Blas and my favorite, the nose speech from *Cyrano,* and everyone raved—even though they all knew them already by heart—because I brought them from France, and in the French manner! You cannot begin to imagine the fervor—it was almost religious—that reigned whenever France was mentioned . . ."

He stopped and lapsed into thought. How could we have answered him? He, too, had been deceived.

From the north came a faraway rumbling, muffled by distance, accompanied by an orange glow that lit up the far horizon of hills and forests. The Germans were bombing the air force base at Tours. The stairs were soon crowded. Leaning their elbows on the balustrade, snifters of old brandy in their hands, Chabannais and his guests harkened to the approach of war.

"Soon it will be the Loire bridges," said the minister. "We'd better not sit around here much longer. Have our gas tanks been filled?"

That was all the sight meant to him.

I looked at Bertrand, expecting a burst of rage. But his face was turned toward the night, contemplative, watchful, oblivious of what was being said, his whole being focused on what he heard in the distance, a message of war meant for him alone . . .

Two days later, on the afternoon of June 13, Chabannais and his ministry, Flat Top, Lily, Mistouflet, and cousin Louis's typist took to the road again, heading south to "sit around" in Bordeaux with the whole Reynaud government. They also took my father, as well as my mother, who never left his side. My father had no illusions. He knew that no one in the government believed in the *"réduit breton,"* the last hope of a stand in France, and still less in withdrawal to Africa to carry on a war already lost.

"Words, words," he said, "political will is dead." In a few days it would all be over. Life would begin again. And meantime, Bertrand and I would be better off in the country than shunted around in chaos. It would be better for our health, moral as well as physical. We were therefore entrusted to our aunts.

So when La Celle and the neighboring châteaux emptied, when the rear of the last limousine disappeared around the bend in the drive, Bertrand simply said, "Thank God!"

My aunt threw open every door and window "to let some air in." After a quick tour of the rooms she said, "They've left me a pigsty!" She sent me to help the gardener clean the lawns, which were littered with cigarette butts, empty bottles, paper, and discarded glasses. Overflowing ashtrays in the bedrooms, the remains of meals scattered about everywhere, dirty plates that had not even been stacked, more glasses and bottles, stains on carpets and furniture, beds unmade. In the bathrooms, sinks and bathtubs filthy, hand towels lying in damp piles on the floor, linen closets plundered . . . "But who were these people?" my aunt said in a shocked voice. It took two whole days to put the house back in order. Scrubbing, polishing, washing, and tidying, cook, gardener, and chambermaid slaved from dawn to dusk. Even Aunt Melly lent a hand, saving Chabannais's room for herself since it was in "such a state" that she refused to "dishonor" her servants by forcing them to work in it.

"And what would that make them think about the people who govern them!" she said.

In my opinion, the harm had already been done. When the last tiny trace of Chabannais's passage had been erased — we even polished the telephone, which my aunt found "greasy," and the doorknobs, one by one — my aunt had champagne served for herself and her servants, Bertrand, my cousin Louis, and me. The vases were overflowing with flowers. Dressed, her face made up, hair done, with nails freshly polished, as she raised her glass you would have sworn she was reconsecrating a desecrated church.

And as that cheerful purification ritual ended, the radio announced (a day late) that German troops had entered Paris without firing a shot. There was no martial music this time. The *"Marseillaise,"* like our government in Bordeaux, had deserted the airwaves. The pathetic dwarf had been struck dumb.

ONE HUNDRED AND twenty miles as the crow flies still lay between Bertrand Carré and Lieutenant Frantz von Pikkendorff. I have already mentioned Frantz's campaign log. It seems to me that he most often wrote it during rest periods or in the evenings, since he took obvious pleasure in making his entries in French. Here is what he wrote on June 14:

Standing in the open turret of my tank, with the Paris wind caressing my face, I gaze up the deserted Champs-Elysées. All you can hear is German military music, crashing out from the Place de la Concorde to the Arc de Triomphe, the roaring of our escort's motorcycle engines, and the clank of tank treads on the road. General Hoth has named my platoon to represent the Fifth Panzer Division in this first victory parade, an honor I would rather have declined and that leaves me with a bitter heart. I know the Champs-Elysées. Hardly two years ago . . . Fouquet's sidewalk café . . . Béatrice, my young French cousin, my mother's niece, asking me with big, naive eyes and with

111

sadness in her voice, "Frantz! Would you make war on us, Frantz?" The world's most beautiful avenue. A symbol. I would have preferred it devastated, its luxurious facades smashed in, smoking, a stink of death floating over the ruins, proving at least that it had been fought for and that it could not be trumpeted to the world that the French capital had gone down without a fight. If that were so I would be happy to parade through it as a conqueror, since the two parts of me, the French and the German, would share the same honor. Instead of which the breathtaking setting is still intact, there are a few cafés open, and soldiers in field-drab uniforms are sitting at the tables . . . Kleist and Ludolf, my tank commanders, smile smiles of triumph at me, and I force myself to respond, hiding my own bewilderment so as not to spoil their fun. Kleist, his Leica glued to his eye, takes shot after shot of the famous avenue and of the infantry massed down each side, ready to parade in their turn behind helmeted officers on white horses. My NCO isn't the only one taking pictures. All the press photographers in Paris — German, American, Swiss, Japanese, Italian — all wearing armbands, have converged on the Champs-Elysées. With the Arc de Triomphe as a backdrop, their photos will be seen by the whole world, proof of France's humiliation.

I halt my platoon at the Place de la Concorde as ordered and from there walk to the Crillon Hotel, where the swastika has flown since dawn. General von Kuchler, Eighteenth Army commander, has invited all section chiefs who participated in the parade to drink champagne with him. An NCO steward in a white tunic barks out our names as we enter the formal rooms. When my turn comes, I am surprised to see the general's aide-de-camp head directly through the crowd toward me, an obscure tank lieutenant.

"Lieutenant von Pikkendorff, the general would like to see you."

"First, a question," says the general. "I'm told that you are twenty years old. Is that correct?"

"Nineteen years and ten months, sir."

The general glances at a card.

"Polish campaign, French campaign," he reads out, raising his voice so that everyone can hear. As silence falls I turn red. "Iron Cross First Class, the *Pour le Mérite* . . . My congratulations. Do you know that you are the youngest decorated officer present in Paris today? I therefore intend to confer a particular honor on you. We have a sacred duty to perform in the name of our great elders, those who just missed the victory we are celebrating today. For twenty years the standards of some of our most renowned imperial regiments have languished at Les Invalides. We will return them to our fatherland. This will be our first priority, and I am entrusting the mission in the name of the German army to you and to your platoon. I have informed General Dentz, until last night the military governor of Paris"—the officers around us laugh—"that he is to be at Les Invalides in person this morning to receive you and surrender the flags. Be correct but firm. Use force if you must, but I don't believe it will be necessary. Here are your orders. Thank you. Your brothers-in-arms envy you."

Horrified. I am horrified. I force myself to think quickly. In three seconds it will be too late. The general has put an end to the interview. He is expecting the regulation "At your orders, sir!" and a sharp click of my heels. Not: "But my mother was French, sir; I cannot accept this mission and I respectfully request that you relieve me of the honor." Yet that is what I want to answer. It's the only reply that reaches my lips from my heart. Fight the French, fair enough. That had been my decision, since I am German and an officer by my own choice. To have beaten them is a part of the fortunes of war. But to humiliate them would be a personal act, and an act of self-humiliation. All eyes are on me, friendly eyes, brotherly eyes; the General is watching me, a little bit intrigued by my silence. So I tell myself, almost in panic, that it's the French who are forcing me to go through this. All they had to do was win, after all! Or blow themselves up at Les Invalides with their own flags and with ours!

Then I hear myself shout, "At your orders, sir!"

They have put an infantry assault section and two open trucks at my disposal. My five vehicles will cross the Concorde bridge in file and turn right in front of the Chambre des Députés. Every bridge in Paris is intact. The public buildings as well, the palaces, the monuments, the Hôtel de Lassay, the Quai d'Orsay . . . It makes me happy and sad at once. Between beauty and honor, the French have chosen beauty. I want to believe that the choice was deliberate and spurred only by concern for beauty, but at times I am not so sure. I don't need

a guide. After all, this is not unlike a homecoming for me. I even
know where our old battle standards have been stored: in a ground-
floor room in the right wing of Les Invalides. I have often visited the
army museum on my trips to Paris. By the river, a few passersby are
watching us. A pretty girl I avoid smiling at, not knowing how it
would be taken. A policeman salutes me, raising his hand to his visor.
I hadn't expected it of him.

The outer gates of Les Invalides are open. The old cannons lined
up on the edge of the moat are mere props. This whole city is a prop
now. General von Kuchler was courteous: my orders are drafted in
two languages, German and French. Courteous but demanding. General Dentz must march in person, followed by our standards, to my
command vehicle, not wait in his office for me or send a subordinate
to greet me. I hand the orders to a guard and bring my small force
into position. My three tanks are lined up straight as an arrow, guns
trained on the elegant old facade, my infantry section in an open
square, and myself, standing in the middle of the square — quite a
sight! The soldiers' eyes shine with curiosity and pleasure. But I feel
uncomfortable. I concentrate on presenting an impenetrable mask to
my men. I console myself with the notion that since we are surrounded by props, well, then, this is theater.

An officer steps forward. Alone. He is a lieutenant. We salute each
other. I wonder where he spent the war. He looks as though he just
stepped out of a nightclub.

"What you have requested is ready," he tells me. "General Dentz
will receive you now."

He says it in German, which he seems to speak reasonably well. I
reply in German. After all, that is the language of the victor.

"My orders are to accept delivery of the standards from the general
in person, on this very spot. Please prepare a table for the inventory.
With a velvet cloth," I add.

Theater . . . The lieutenant swallows and makes an about-face
without pressing the issue. Well, at least he made an effort, however
weak. I have no pity for him. If he had looked like a fighting man I
would not have demanded the cloth, an idea that came to me suddenly. In five minutes the table appears, complete with a superb
velvet cloth of midnight blue. Only the general is missing. Will I hear

a far-off detonation announcing that he has blown his brains out, which is what I would do if I were a German general required to come out and kiss the boots of some paltry French lieutenant? Ah, no. Here he is, accompanied by a few officers. He is pale.

I hear him exclaim as he sees me, "And on top of that, he's just a kid!"

That remark consoles me as well: now it's not Germany's revenge over France, but youth's revenge over old age. Police officers now appear, carrying armloads of flags wrapped tightly around their staffs. The whole proceeding lacks dash and style. How old these Frenchmen are! They don't know how to play anymore.

"These are not broomsticks," I say harshly to the lieutenant, who translates. "Present them to me unfurled, with respect, if you please."

I stand my troops to attention and proceed with the inventory, the list in my hand. Several Imperial Guard battle flags, some from the Royal Bavarian Guard, the Death's Head Hussars, the Empress's Dragoons, the Saxon Lancers . . . October and November 1918. My father had told me of those months of agony. Regiments reduced to twenty men, almost all of them officers, starving, exhausted, despair in their hearts . . . General Dentz affects indifference, ostentatiously chatting with his officers. This deserves a lesson. When they are about to load the flags onto my trucks, I give the order to present arms. Upright in their turrets, as rigid as if on parade, Kleist and Ludolf salute. They too, are twenty years old. I owe it to them to play my part well. I set a cat among the conquered pigeons.

"That order applies to the French general and the French officers as well!"

This time, I use my etymologically maternal language, which I speak with absolutely no accent. Surprise freezes them in their tracks. They stare at me with alarm, they salute, the general last, after a moment's hesitation. Had he done it instantly, had he shown himself equal to the emotions of that moment, since the scene seemed made for drama I would have gone up to him and thanked him, confessed my French origins, offered him my sympathy, and apologized man to man. The two of us would have been magnificent. I would have had tears in my eyes, I know myself . . . But since that was not the case, I hold it against him twice over, German to Frenchman, and take my revenge.

Still in French, I spit out the words, "With the kid's compliments, sir!"

He gives a start, mumbles something I don't catch, turns on his heel and leaves. He botched his exit. The play was superb, but it took place without him . . .*

I'm staying at the Crillon Hotel as General von Kuchler's personal guest. I take long walks in the city. The metro is running. All is calm. Many cafés and stores are open; people seem to be adapting to our presence. Our soldiers stroll about like tourists. They buy postcards. They take photographs of one another in front of Notre Dame or at the base of the Eiffel Tower. At the Tomb of the Unknown Soldier they remove their caps and stare. It is precisely this sense of almost blatant normality that bothers me, even if some Parisians pretend not to see us in the street. The German army peacefully visits Paris, politely buying silk stockings and drinking coffee at sidewalk cafés! If I were myself a Parisian, it seems to me that I would feel humiliated, as though I had been deprived of a war for which I had been judged unworthy because I had not been up to playing with the grown-ups.

I called Béatrice's number from a café phone box. I let it ring for a long while, without success, and was simultaneously disappointed and relieved. There is no reason to believe she would be pleased to see me arrive at her home in a German officer's uniform, and I have no civilian clothes. In any case, regulations forbid it. By the end of the day, however, I can no longer contain myself. Béatrice lives on the rue Franklin, not far from the Trocadéro. From her balcony there is a splendid view of the Seine and the École Militaire. Her parents have often entertained me there, although last year, after Munich, with a certain reticence. Her mother was my mother's first cousin. We had exchanged a few furtive kisses on the balcony and thought sadly of everything separating us. I don't think we were truly in love. We were

---

*General Dentz met a tragic fate that contradicts the low impression Lieutenant von Pikkendorff had of him. Stubbornly loyal to the Vichy government, he commanded French troops in Syria and fought courageously against the British and Free French. Imprisoned in 1945 and condemned to death, then pardoned but held in solitary confinement, he died of sadness, neglect, and loneliness, his legs in irons.

acting out a Corneille tragedy and all its conflicts, so fascinating to young lovers these days. The conflict of love and honor. The romance of impossible loves . . .

On rue Franklin the concierge recognizes me. My uniform leaves her speechless. I believe that from now on she will be a staunch believer in the Fifth Column stories the French so love to frighten themselves with. She stammers, "Oh me, oh my, oh . . ." as she backs toward the other side of her room like a trapped animal. Finally she calms down: "No, there's no one here. They all left a week ago, Miss Béatrice and her parents . . ." Ignoring the wheezy old elevator, I take the stairs to the sixth floor four at a time. The bell doesn't work. I knock again and again on the double door without any hope of a reply. That doesn't prevent me from pressing my ear against the door to listen, heart beating, for the footsteps of someone coming to open the door. I stay there a full quarter hour, motionless on the threshold of my French past. I miss Béatrice. We could have played one last heart-wrenching scene together . . .

The German High Command has authorized the reopening of the Pigalle nightclubs. Some of them had never closed down and were still lining up their regiments of dancers completely nude. At least these troops haven't surrendered but march resolutely on to combat. My feet carry me to *La Nouvelle Eve*. It's packed with German officers. Many of them had seen me with the general at the Crillon and ask me to join them, but I prefer to be alone. I choose a little table apart from the rest. Béatrice would never have agreed to come into such a place with me, yet here she is on the plush velvet banquette beside me, in fact pressed tight against me, in the shape of a beautiful woman who a moment ago was dancing nude on the stage and who now is dressed, although it hardly makes any difference, and has invited herself to sit at my table. I have drunk a lot, naturally. I'm beginning to feel a little drunk. With a hand slipped under the strap of her dress, I tease her nipple and call her Béatrice. The darkness helps.

I cover her with kisses, and I ask her, "Answer me honestly, Béatrice. Could you love a German officer?"

"This very night, my darling, if you want," she tells me, rubbing against me.

It isn't Béatrice's voice. That sobers me up. I leave.

A cloud of cigar smoke fogs the Crillon bar. It's just as packed with officers. I finish myself off, sadly, with cognac. There really are too many Germans in Paris. My God! What has happened to the French? What has happened to us? Is there even one to redeem the rest?

I rejoin my regiment tomorrow, at dawn, at the Porte d'Orléans. We will head toward the Loire. It seems that the war isn't yet over, even if it now seems to be turning, as General von Kuchler said yesterday, into a peaceful occupation of the whole of France . . .

THE LAST FIGURE even vaguely representing organization to cross our path was the Petit-Bossay mailman. In those days rural letter carriers wore an old-fashioned uniform halfway between that of a customs officer and a forest ranger, which conferred on them a borrowed authority. Our mailman was pedaling along the road to Beausoleil, which ran close to Blue Island, with a telegram addressed to the "Bonnadieu Family." It was June 17.

"Bad news!" he shouted, stopping his bicycle for a moment. "The Châtillon police should be doing this, but everything is head over heels right now. I'm replacing them. If I were you, I'd go warn your aunts."

By bicycle, by car, everyone was soon gathered at Beausoleil around my aunt Germaine Bonnadieu, who sobbed silently, her face contorted. The telegram passed from hand to hand: "The Ministry of War regrets to inform you Lieutenant Raymond Bonnadieu Sixth Chasseurs Regiment killed in action Arras sector May 16 stop Deepest regrets."

"A whole month!" my aunt Germaine murmured. "Every day I begged God to return him alive and he had been dead for a month . . . And how can I tell his father? Is *he* still alive?"

Have no fears for my Uncle Léonce, commander of a mounted artillery battery and emperor of the Beausoleil hunt. Taken prisoner at Bourg-en-Bresse with ammunition and supplies intact and without firing a shot, he was at that very moment standing within a temporary barbed-wire enclosure and planning a monumental and ruinously expensive work on woodcock shooting, to be published after the war at the author's expense, a work that five years of studious and serenely borne captivity in Germany would allow him to accomplish nicely . . .

My cousin Raymond had, for a month, lived something like a posthumous existence in the hearts of those who loved him. He thus died twice instead of once. This made a deep impression on my aunts. It was now that Aunt Germaine started to lose her grip. "This telegram," she said, "might never have arrived . . ." Then she crumpled it into a ball, swallowed it, and declared that it never *had* arrived. She continued to pray every day as though her son were alive. It was no longer possible to leave her alone for long in that big house, which she refused to leave so as "to be here when Raymond gets back." My aunts took turns staying with her, and "poor Germaine's condition" became their greatest preoccupation.

Increased freedom for our little gang followed. "We must trust Bertrand," Aunt Melly said again, adding only one condition, that he come home with me each evening, which was ritually translated into a glass of lemonade with my aunt in the garden (while I was asked to go and wash my hands, change my clothes, and clean my room), or into an invitation to stay for dinner when the Majorels were coming to join us. Knowing that we were on Blue Island, as far from the war as one still could be, the aunts distractedly left us to ourselves. Even Aunt Octavie de Réfort stopped trying to prevent Maïté from spending all her time with us.

Noting the change, Bertrand coldly remarked, "At least your cousin didn't die in vain!"

And indeed we did owe him our extra freedom. As for the use Bertrand would make of it and why it so gladdened him, it is vital at this point to suspend judgment, for otherwise his remark would be tainted with cynicism and a spiritual ugliness that were not part of his character.

Until that June 17, the inherited harmony of our childhood had endured. Nothing so far had cut us off irrevocably from La Celle, La Guichardière, Beausoleil, La Cornetterie, La Jouvenière, the village of Petit-Bossay, the aunts, the uncles, their families, their friends—all the people and things we saw every day, our every moment's universe . . . And if the unbroken succession of national disasters we heard of seemed to us to to be somewhat their responsibility, it was only because of their age, not because they had fallen from grace in our eyes. In fact, the interlude of the government in full retreat had brought them closer to us in a communion of scorn. They were still our kind, as opposed to "them." After June 17, though, everything changed, and we found ourselves alone; that is to say Bertand found himself alone and dragged all of us, Maïté and me, Pierrot, Zigomar, and Zazanne, into the underground recesses of his solitude.

First came Pétain's speech. We listened to it on the evening of the seventeenth, frozen around the radio in Aunt Melly's living room with the Majorels and the Réforts. The old marshal had just replaced the dwarf at the head of the Bordeaux government. In Chabannais's choice phrase, he had indeed been left holding the baby: the armistice. For me, as for every Frenchman who lived through those days and heard that speech under those circumstances—with the army and the nation in naked flight— its tone as much as its words have been branded on us as if with a hot iron. His quavering voice, broken by emotion, the voice of an eighty-four-year-old, and (even after so many of the dwarf's evasions!) a sincere voice, a voice bound by ties of blood to the country's misfortunes . . . Everyone has heard the famous lines: "Certain of the affection of our admirable army, which fights on with a heroism worthy of its time-honored military traditions against an enemy superior in number and in arms . . . I offer France the gift of my person to ease her misery . . . . In these painful hours, my thoughts go out to the miserable refugees who travel our roads in extremes of destitution . . . It is with heavy heart that I tell you today that we must cease fighting. This night I have addressed myself to the adversary to ask if he is ready to seek with us, between soldiers, after the battle and with honor, the means to put an end to hostilities. . . ."

A requiem.

The older I get, advancing like everyone else along the primrose path of French decadence, the more I realize those words should never have been spoken. In exchange for a brief respite, material rather than moral, they brought nothing but abnegation, illusions, and divisions, in which the better part of the nation's elite and of what is called traditional France were long mired and whose price we are still paying . . . But that too is another story, and my own personal weaknesses should deter me from dwelling on it.

Of course I did not in the least feel all this at the time. Surprise was my overriding emotion. Not the surprise the old man's words might have caused in me (their true meaning escaped me), but the surprise — the stupor — I felt at seeing the faces of my uncles and aunts break down under the corrosive influence of the words. Closed within their own coffins, they were hearing the fall of the earth from the grave diggers' spades. Uncle Armand looked as though he were crucified, his gaze focused inward, dying in the bloom of health. Aunt Sophie wept unrestrainedly, slumped deep in her armchair as though suddenly an old woman. Slumped over too was Uncle Gaetan de Réfort, a man who loved to thrust his chin forward to emphasize his superiority. He held Aunt Octavie's hand, his eyes vacant, looking vaguely relieved as if he had crossed the threshold from life into death. My memories exaggerate the picture, I am sure, but this was indeed the impression left me: that they had aged a hundred years. They were without spirit. They had left us.

Bertrand breathed into my ear, "All the same, it's not the end of the world. What's got into them?"

The marshal had finished speaking. The *"Marseillaise,"* naturally, yet again slaking with base blood the thirst of our French fields . . . Someone turned off the radio.

In the heavy silence that followed, Uncle Gaetan found strength to say, "God be praised! At least this means that young lives will be spared . . ."

Bertrand grabbed my wrist.

"And why should they be spared, answer me that. Tell me what will they do with their lives? They'll drag on . . ."

I heard Maïté murmur, "Bertrand . . ." It wasn't a protest so much as a token of loyalty. Their hands met in the dark corner of the living room where we had taken refuge. I know that eyes do not gleam in shadows and that it is a bad literary convention to make use of that hackneyed image. Nonetheless it is the only one that fits my memory of that scene. Aunt Melly watched them, and her eyes shone in the shadows . . .

So it was that "they" finally disappeared. Soon they no longer mattered to us. Three days later, in fact, on June 20, Uncle Gaetan de Réfort took it upon himsef to sweep away what remained of our illusions, sweeping himself away into the bargain. It was Pierrot who came to get us on Blue Island. In accordance with Bertrand's "new order of battle," he had been acting as our "border guard," a task that consisted of patrolling the area between Petit-Bossay and the main road on his bicycle and watching the movements of the world from which we had withdrawn while awaiting "the enemy." But this wasn't the enemy. This was the French army. At least, what was left of it.

"They've set up machine guns at the entrance to the village!" Pierrot panted. "They've requisitioned wagons and blocked the road! The policeman has gone to find the mayor."

Uncle Gaetan was the mayor, as all his forebears had been, and general councillor for the Châtillon district into the bargain. When we reached the village — our whole group, including his daughter Maïté — he was in the middle of a shouting argument with a young captain, and he seemed to be getting the worst of it.

It must have been three in the afternoon. Taking advantage of
the farm buildings that formed a bottleneck on the outskirts of
the village, the "French army" had thrown a defensive block
across the road. Our own "fortifications" on Blue Island could
hardly have been more ludicrous. They had pierced rough
loopholes in the sides of old barns, piled up carts and wagons to
form a crude barricade, and concealed a small howitzer beneath
a covered gateway. A score of helmeted soldiers armed with
submachine guns stood perched on the improvised barricade
like the tattered mannequins of our Blue Island garrison . . .
No matter, there was heroism in the air. A small triangular pennant fluttered gallantly above them, mauve and green with
two crossed rifles ringed with silver letters: First Squadron,
Eighteenth Regiment of Dragoons. Standing with legs apart,
battle-dress tunic in tatters, shirt open and chest bare, glorious,
indifferent to the pacifist entreaties reaching him from the barricade where the last of the refugee flow was trying to squeeze by, a
beardless NCO scanned the road ahead with binoculars.

"Come now, Captain," my uncle was clucking, "can't you see
you're in the way?"

"In whose way? In what's way? I'm fighting a war! Hard to
fight a war without getting in civilians' way, Mr. Mayor. And if it's
them you're worried about, you can stop worrying. In three
hours not even a cat will be coming through here. The stream is
about to dry up, Mr. Mayor."

We had noticed that ourselves in the course of our "border
guard" missions. The trickle of runaways was not comparable to
the crush of a few days earlier. Exhausted pedestrians pushing
baby carriages or overloaded bicycles they no longer had the
strength to ride, the residue of the exodus, the poorest, the least
resourceful . . . Every now and then they would shout insults at
the soldiers blocking the road.

"The war is over, you fools! A little late to be playing soldiers."

Since Pétain's speech they had indeed believed the war to be
over — and here it was still going on, just because of a few
dangerous madmen. It meant that they had to keep walking.

"They're out of luck, those poor bastards," the captain commented coldly. "If they'd dawdled just a bit longer the Germans would have overtaken them and they'd be going back home now. Well, they're the last. After that it'll be quiet for a while, and then the Germans will come! Perhaps this evening. Or tomorrow morning!"

He said it almost cheerfully (Uncle Gaetan's pince-nez dropped from his nose when he heard it), as if it were the best news of the whole day, war at long last, glorious war! A small, thin man with a Musketeer mustache and a week's growth of beard, his left arm in a sling, the empty sleeve of his tunic stained dark with blood, clearly exhausted, staying on his feet through willpower alone, willpower and the pleasure he felt in thumbing his nose one last time at fate. . . He laughed.

"You think it's funny?" my uncle asked, taken aback.

"You don't?"

This time Uncle Gaetan exploded. I knew that that kind of anger masked intense fear.

"You know perfectly well that it's all over! You heard the marshal? He told you to stop fighting. What is all this vainglory, and who do you think you are? Is some jumped-up little captain going to teach Pétain a lesson about honor? The Germans are in Besançon, Lyon, La Rochelle, and you want to stop them here! All by yourself! Do you want to see this village destroyed, this village that never asked anything of you and that only wants to be left in peace? As mayor of this village and general councillor, I order you either to clear out or to surrender to the German authorities as soon as they arrive."

My uncle was trembling with emotion. He thought he was sublime. The savior of the village. There were a few peasants with him, Magloire the grocer, the policeman, the mailman, the clerk, the curate. They all solemnly approved.

Bertrand poked me in the ribs and said with a poker face, "Your uncle is offering us the gift of his person."

"Surrender?" said the captain. "Clear out?"

He had obviously taken it badly.

My uncle must have noticed. He began again in conciliatory tones, "If this is a question of honor, Captain, I understand very well. I suggest that you avoid opening fire and withdraw immediately. Report to your superiors that your machine guns are jammed. That way appearances will be saved."

He seemed delighted with his solution.

"Sparks are going to fly," Bertrand said to me.

We were standing to one side, attentive but already detached. Searching through my memories I find no particular emotion. We simply counted the points scored on either side. The outcome no longer concerned us. Maïté listened to her father as though he were a stranger, her eyes void of expression. Yet we liked this captain. He had courage and style. Deep down, we pitied him. We knew intuitively that whatever he did, whatever he said, "they" would win. "They" could not stomach the fact that he was different from them and that he persisted in demonstrating the difference . . .

And sparks did begin to fly. The captain grabbed Uncle Gaetan by the lapels of his hunting jacket.

"I don't take orders from you! I don't take anyone's orders anymore! I know that they've asked for an armistice, but the armistice hasn't been signed yet.* I intend to wait for it here, facing the enemy. What is this hamlet of yours called?"

"Petit-Bossay," my uncle wheezed.

"All right then, Petit-Bossay is France! I'm not moving from here. This village will be defended. This is all that remains of France and of my regiment. We are saving honor, Mr. Mayor, which you call appearances . . ."

The captain released my uncle. His tone changed. It was with something like despair that he went on, "Honor, Mr. Mayor, can you understand that?"

He was controlling himself with great difficulty. We sensed that he was on the verge of tears, about to break from nervous strain.

---

*It was not signed until June 25.

What happened next was particularly pathetic. Uncle Gaetan seized the captain's hand and for long moments held it in his as he addressed the man in a penetrating vibrato, tearful, compassionate: "But of course we understand. We, too, are torn. Just circumstances, isn't that right? Circumstances, Captain . . ."

Then everyone spoke his heartfelt little piece. A few even essayed a tear. Honor was being buried: condolences rained down like earth onto a coffin, sympathy overflowed along with admiration for the deceased (although it had to be acknowledged that he hadn't really kept up with the times) . . .

Even the curate came out with a homily that summoned Christian charity to help them cope with the funk gnawing at their insides: "Think about the mothers of these young soldiers, or their sweethearts. It would be no dishonor in these circumstances to think of them . . ."

Haggard, buffeted, trapped, the poor captain looked as if he was about to apologize for still being there.

"So?" Uncle Gaetan asked, sure he had carried the day. "The right decision, this time?"

The captain kicked vigorously free of the insalubrious quicksand.

"Get the hell away from me!" he said.

"There'll be a second round," Bertrand commented, "but the champ looks dazed. I'm betting on Uncle Gaetan."

"So am I," said Maïté in a neutral voice.

It didn't last long. Less than an hour. My uncle returned to the fray. He rounded up the whole village, particularly the women — "mothers, sisters, and wives of soldiers!" — and stressed the gravity of his intentions by donning his mayoral sash. With the three glorious colors striping his fat stomach he advanced, flanked by his municipal council (which included veterans of the last war, medals and Croix de Guerre to the fore). They could not have made a braver showing if they had come with the intention of exhorting the captain to heroism, even at the risk of their own lives and belongings.

"Captain," my uncle began, "I represent the only legal authority here. It is my duty to inform you that if you do not obey my order to withdraw, I will myself go to the Germans and attempt with their help to negotiate some way of sparing the village and to arrange some breathing space for you to abandon your positions and escape being taken prisoner. I appeal to your intelligence and to your sense of responsibility."

The captain's weary eyes traveled over the faces that surrounded him.

A woman gave a vindictive yell, "Don't you think you men have done enough already?"

She was certainly not a bad woman. Circumstances . . .

"Mr. Mayor," said the captain in a flat voice, "at least you have taught me something. That the soldier's true enemy is not the one he faces but the one who stands at his back, ready to betray him."

"How dare you . . ." my uncle began, white as a sheet.

"And so," the captain went on, waving the interruption aside, "I will not hesitate to open fire on anyone who takes it upon himself to leave this village with a white flag. Now, go to hell!"

At that precise moment the loud noise of engines reached us from the road and grew louder second by second. In reality, perhaps, it happened a little later, but this shortcut in my remembering seems to me to convey the drama of the scene more effectively. The Germans! we all thought. The municipal council fell flat on its collective face.

The NCO with the binoculars reported, "It's not coming from there, Captain. We've been outflanked!"

"We" had indeed been outflanked. By the French. A whole procession of limousines preceded by motorcyclists, who came to a halt at the barricade with a great screeching of brakes.

"Negotiators! Negotiators!" one of them cried.

The cars were crammed with uniforms, a great many stars and stripes on the sleeves. There were also some civilians with faces as somber as their black Sunday suits and waistcoats.

"The white flag," Bertrand said calmly. "Now would be a good time to fire on it, wouldn't you say?"

Each car had its own white flag, and the lead car had two, as well as a starred tricolor pennant. A general in a red cap stepped out. I have checked the records: it was General Huntziger, leader of the French armistice delegation.

"Who is in command here?" he asked.

"I am, sir," said the captain, approaching the general almost casually.

"Astonishing," said the general. "We thought we'd find the Germans here. We thought they were already at Poitiers."

"Well then," said the captain phlegmatically, "they must be late, sir. Unless it's you who are early . . ."

He made a small gesture of apology. It probably meant that if he permitted himself such remarks, it was to keep from crying. The general accepted it.

"Good. What about the road?"

"Almost no one left, sir. Within an hour it'll be deserted. In my opinion, you'll find the Germans at Loches."

This time I was the witty one: "At The Three Ducks," I whispered into Bertrand's ear. "Uncle Sébastien will be pleased . . ."

"And French troops?"

"If you mean fighting troops, I haven't seen anyone behind me for two days," the captain answered.

Uncle Gaetan walked over and whispered a few words to General Huntziger.

The general coughed as though clearing his throat, and came to a decision: "I am told you intend to resist, Captain . . ."

"Quite right, sir."

"I cannot reproach you for that, nor can I order you not to. Who could give such an order, anyway? Do you know where the General Staff is at this moment? At Montauban. Would you have guessed it? All I ask is that you think about it."

With white armbands around their sleeves, generals were wandering around the cars, saying nothing, avoiding the gaze of the dragoons still at their posts on the barricade. An admiral with prostate problems urinated against a wall. Ostentatiously consulting his watch, a colonel observed that they were falling behind schedule.

"We're here to relieve you," the general went on with sad humor—his features were strained by fatigue and lack of sleep. "The French armistice delegation is on its way to the front! You've done more than your duty. What is it to be?"

The captain gestured hopelessly. These weren't the words he'd hoped for. Everything was giving way. Things were falling apart all around him. Generals with white armbands, peasants eaten up by fear . . .

"In any case, we have no more gas."

"Then walk," said the general. "Go on foot. Stay alive and free. Have the courage to walk. I am asking you . . ."

Those remarkable words are still in my ears. His tone was almost gentle, unsoldierly, affectionate, that of a father to his children. There was something pathetic about this beaten warlord, faced with the last fighting men he was to meet before crossing into the German lines, wanting only to save them.

"You've just learned how to flip a pancake," Bertrand said to me, affecting complete disdain.

Beyond the fact that this struck me as needlessly unfair, his tone intrigued me: it was as though he was relieved, delivered from some ill-defined fear. Only now do I understand why. Picture a different climax to this scene: the "pancake" handing the nice fatherly general his walking papers; Petit-Bossay pretending to be Blue Island; and Bertrand relegated to understudy for the one role he had chosen to star in but which someone else had had the gall to usurp . . . His pride would never have stood it. He would rather have given it up. Had he been forced to plagiarize there would have been no more chance of outdoing himself. Instead of which, fate had returned to its original course: an organized dispersion, Huntziger's funereal convoy on one side, the captain and his dragoons on the other.

For the captain gave up without another word. A shrug sufficed to summarize his thoughts. The barricade was dismantled. The peasants pushed and shoved at the wagon wheels. Among them was Pierrot's father, throwing us furious looks: "Couldn't you lend a hand, lazy bastards!" We did, with our fingertips. The howitzer was tossed into a pond.

"Faster! Faster!" bawled Uncle Gaetan, as if expecting the Germans to come into sight at any second.

Looking like a dirty old man offering candy to children, the curate handed out coffee and sandwiches to the dragoons, who had been offered nothing to fill their empty stomachs since their arrival. Making sure they would have the strength to leave . . .

They moved out in single file down the road, alongside the last refugees. Soon you could no longer see them clearly. From behind, the two groups were indistinguishable.

Just as quickly, the village emptied. Everyone went back to the shelter of his home and slammed the shutters closed. They looked none too proud as they left, each of them shaking Uncle Gaetan's hand as though someone in his family had just been buried. In life there are cruel coincidences, so cruel that they can't be coincidences—they reveal fate's deliberate intentions. That same day, June 20, at the same hour, officer-cadet Jean-Louis de Réfort, Maïté's brother and Uncle Gaetan's son, twenty years old, met his death defending one of the Saumur bridges, between the north bank of the Loire and Doffard Island, today called Cadets' Bridge. We didn't hear of it until ten days later, and I have never been told how my uncle took the blow. I don't know whether he found in it the answer to the question framed by the captain at the foot of his barricade: "Honor, Mr. Mayor, can you understand that?"

As at the end of every afternoon, the time had come for us all to go our respective ways. We stared at our feet in silence, avoiding looking at Bertrand and asking him *the* question that was burning our lips. It was Zigomar, painfully swallowing, who took the leap.

"So you think it'll be tomorrow?"

"Tomorrow or never," said Bertrand.

"Are you still determined?"

"Why would I have changed? You saw it, we're the only ones left . . ."

Pierrot did his best. "But what about the armistice?"

"Did I sue for an armistice? Are you afraid?"

"No," said Pierrot. "But it's my father. You saw how he looked a minute ago. He'll keep me in the house . . ."

"You're afraid," Bertrand repeated. "And Zigomar?"

"Well, it's my mother," said Zigomar.

"My compliments," said Bertrand. "You were braver in the bus last year when the girls were pulling their pants down. Well, what are you waiting for?"

Sheepish, blushing, we looked of course at Maïté. She stared back at us. I have tried to decipher the meaning of this memory. I can still see Maïté in the flickering light of the old bus, with her dazzling white skin, her pink nipples, the light blonde shadow like a sacred triangle beneath her belly. A pact. A carnal pact, a chivalric rite, something that had bound us and from which there was no release, a communion in the mystery of life, of the flesh, of death. All very vaguely perceived, I remember, expressed in silences, in unconscious impulses of the soul — but what I also remember is that now this same sense of communion swept our hesitations aside as soon as Maïté turned her faraway gray eyes upon us.

"I'll be there," said Pierrot. "I promise I'll try to come."

"Me, too," said Zigomar.

"And you?"

The question was addressed to me. Bertrand wasn't just asking for form's sake. He expected a clear answer. And here again, I have never really deciphered the reason for his insistence. Either he was giving me a last chance by pretending to trust me, or else, knowing in advance that I would break, he was condemning me to guilt . . .

I at once hid behind my cumbersome personal camouflage, fooling no one. I borrowed my style from *The Three Musketeers*: "At your orders, captain!" I bellowed, clicking my heels and raising my hand to an imaginary visor, my eyes warlike.

Pathetic, of course . . . I felt Maïté's eye leave me.

"At ease," said Bertrand. He was mocking me. Then, more seriously: "Tonight you'll stand guard with me."

"Tonight?" I stammered.

My legs wobbled beneath me. My heart hammered in my chest, and to make things worse I was terrified that my fear would be discovered.

"Well yes, tonight. What if the Germans come through tonight and we're all at home asleep? They'd find Blue Island empty. They wouldn't even know that it belonged to me. They would cross it without even noticing it. It would be as though it had never existed, don't you understand? As though you and I" — but of course he was thinking of himself alone — "as though we had never existed! Never!"

That was Bertrand inspired. The Bertrand of the mad tirades, the only son of Squadron Leader Carré at his post on the Chinese border, and of Laïcha, Oriental dancer, who had given her son to France, prince of the fabulous bracelet and of ghosts going bump in the night. There must be many explanations for the theme he had chosen. I pondered the question deeply during my convalescence, and now I wonder whether, save for a Blue Island somewhere inside me, I have ever truly existed.

He added, "Kolb has given me his gun."

Which made us jump. Decoded, this meant: "From now on I'm playing in earnest." At least we had been warned, and I am sure he said it with that intention and at exactly the right moment. For me it was the last straw. I tried to get out of it.

"Tomorrow, all right. But tonight, no. What if my aunt found out? She'd keep me in, too."

"She'll never find out, idiot. You just have to wait until she's asleep. I'll do the same at my aunt Sophie's, and then I'll come over for you. Just leave your window open. You'll get back in just before dawn and pretend to get up. If not, we all meet at eight o'clock tomorrow morning . . ."

I went to bed early. My camouflage had melted like third-rate makeup. I needed to be alone. Aunt Melly gave me an absent-minded good-night kiss. I turned the key in my door and lay fully dressed on the bed, determined not to leave it. Keeping my clothes on was one last camouflage device, one last attempt to fool myself, as if I were near to taking the leap and some mysterious force were keeping me from it. Pitiful . . . I lay in the dark for a long time, eyes open, listening to the sounds of the night. I reeled off all my prayers. O God, deliver me from Bertrand, from the Germans, from girls, from myself . . . I was afraid, afraid of being afraid, afraid of showing that I was afraid, afraid of agreeing to follow Bertrand when the moment came because I knew that I would be afraid, afraid of refusing because I was afraid. Fear beat in my temples and in my heart like the mighty waves of a storm. It was blowing hard outside. A branch fell from a tree with a sound like an explosion. In cases like these I had no other recourse than the Coward's Prayer to stave off total breakdown. My hands folded, I recited it almost with greed. I had made it up myself. Rediscovering the old words of the litany and digging up new ones occupied my mind to the exclusion of everything else: "O God, take pity on cowards for they are the most miserable. O God, take pity on the craven, for they are the most wretched. O God, take pity on the chicken-hearted," and on through the frightened, the timid, the yellow-bellied, the scared, the terrified, the funkers, the fearful, the faint of heart, the nervous, the timorous . . . To this repertoire from the year before, which had come to me during the night spent deep in the Salbris forest when we had cycled to join the Free Corps, I added: "O God, take pity on the wet hens, the cringing, the lily-livered, the pusillanimous, the . . ." I fell asleep exhausted, drained of emotion.

At some point in the night, half-asleep, I heard muffled footsteps in the corridor, and (without wondering who it might be, since Bertrand couldn't be coming from that direction) lost consciousness once again, my soul at rest. Probably he had given up his plans. I don't know how much time went by. A handful of pebbles thrown from outside clattered onto the floor of my room. Fully awake this time, I crept toward the window on my belly so as not to be seen from below and, concealed behind a curtain, saw Bertrand's silhouette in the shadows by the light of the moon. I stayed as still as death. A second handful of pebbles struck an open shutter and fell back to the ground by the side of the house. He waited a few moments, most probably knowing there would be no response. I heard him murmur, "As you

like . . ." Then he mounted his bike and disappeared down the
drive. I looked at my watch. It was three o'clock The hour helped
assuage my guilt. At three o'clock, I thought, with the war going
the way it is, the Germans are probably asleep as well and
Bertrand won't be needing me tonight . . .

I undressed, relieved, and slid blissfully between the sheets,
muttering the coward's rosary and postponing for the morrow
the immense fear staring me in the face.

At the end of that same day, Lieutenant Frantz von Pikkendorff
noted briefly:

> We are making haste slowly, almost bucolically. It is mostly the
> refugees who are holding up our advance. Some of them are even
> beginning to turn back now that war's dark wings have passed over-
> head. We could take side roads parallel to this highway, there is no
> lack of good alternatives, but apparently the High Command has no
> objection to this relative slowness, at least in this sector.* Right in
> front of our noses operations were halted for the whole afternoon
> along a front six miles long and six deep. We had to halt. We swam in
> a beautiful little stream. The reason for the break was the arrival of
> the French armistice delegation just a moment ago. I was the one
> who welcomed it before it continued on its way through German
> lines. Apparently that was yet another honor for me. A French general
> shook my hand. Stooped and very gloomy. He asked for a glass of
> water. In his place, I would have demanded champagne. I would have
> drunk it in swashbuckling style, saying to the German officers pres-
> ent, "My compliments, gentlemen! Well played! Until next time!"
> Then I'd have smashed my glass at my feet . . . It's not really that dif-
> ficult to make a dramatic exit!

---

*This is correct. Even before armistice negotiations began, Hitler had decided to cut
France in two. For political reasons he spared what was to become the so-called free zone
(whose capital would be Vichy) by slowing the German advance in the center, but at the
same time he unleashed his Panzer divisions on the two wings, Bordeaux in the west,
Switzerland in the east.

And it was on the evening of June 20, at The Three Ducks in Loches, that the doughty fifty-year-old Captain Sébastien Lavallée, my uncle, wisely abandoned plans for the village's defense. His "troops," a supply-depot company, had fled without warning him during the night, leaving him alone with his orderly. He surrendered with dignity. At that very hour, a prisoner on parole in his room, he was attacking a pâté en croûte and an endive salad brought up to him from the kitchens, along with a bottle of cool Bourgueil, and listening sadly to the toasts proposed by victorious German officers in the hotel dining room downstairs, accompanied by irritating heel-clicking.

But that, too, is another story.

IT WAS EIGHT when I woke up. The day was already in full swing, and light poured into my room through the open window; I sensed its brightness through my closed eyelids. I knew that awakening would mean facing all the terrors of the night before but that I had to find the courage to take the leap or never respect myself again. So I put off the inevitable. I dragged along like a lost soul in one of those limbos between sleep and wakefulness in which those who dread life grant themselves the illusion of a last reprieve each morning. But such moments of respite cannot be prolonged. Roughly and inexorably, life yanks you from your refuge. You must put on your hand-me-down humanity and pretend to be satisfied with it, present composed features to the world, and leap into the water without knowing how to swim. At the very end of life the effort is just as difficult as it was in the beginning. Some people develop a resistance to drugs, to nic-otine, to alcohol, to what you will — but never to life as it appears on first opening one's eyes on a new morning. It was in this frame

of mind that I got out of bed on that twenty-first day of
June.

Squares of chocolate and slices of buttered bread wolfed down
in the kitchen picked me up a little. Aunt Melly's maid, who
cycled in from Petit-Bossay every morning, was there as usual.
Apparently it did not occur to her that she had performed an act
of courage by taking to the road today, when everything seemed
to indicate that the Germans would be here any minute.
Although not even the smallest sign of their advance units had
been seen, they were already a tangible part of the landscape on
that peaceful spring day. As Frantz had written, not uncharitably,
war's dark wings had passed overhead. According to the maid,
the village and surrounding area were completely calm. Everyone was just "going about his business." And after all, we might
not even see these Germans. They would probably go past
far from here, on the main highway. Why would they waste
their time here when even the French soldiers hadn't waited for
them . . .

That confident view of the day put me right back on my feet.
And when the maid told me that Madam had had to leave early
for Beausoleil in order to take care of poor Madam Germaine —
my aunt Bonnadieu — who was calling out for her as if for the
Lord Jesus Himself, that she would be back for lunch at noon
and would like me to be there as well, and please not to wander
around too much, I felt my heart swell with relief. I was already
halfway to salvation. All told, starting with my aunt's request that
I be at home for lunch (a request I fully intended to honor — nor
did I intend to budge from home after lunch), plus the fact that I
had gotten up late, plus the time I would gain for myself by
taking my leave of Bertrand well before noon, my period of risk
was reduced to three short hours (on the unlikely condition, of
course, that any German might wander into our neck of the
woods). So I was whistling a merry tune as I leaped on my
bicycle, my Confederate cap low on my forehead, and pedaled
off through the sweet-smelling air to "return to my post" on
Blue Island.

Nothing could have been more peaceful than the approaches to the island on that morning of June 21. It seemed inconceivable that the exquisite play of light on water, the murmuring of the river around the old mill, the sweetness of the spring air, and the marriage of the foliage overhead to a sky filled with wispy, sailing clouds could be the background for anything other than a pleasant outing among friends. This almost completed the process of calming my fears. Freed of my worst apprehensions, I dismounted at the bridge and slipped into my role in the most natural — in other words, serious — way in the world.

Cupping my hands, I hailed my friends: "Garrison ahoy!"

Since my last visit a barricade had been erected at the far end of the bridge on the northern tip of the island — five recently felled poplar trunks, hastily stripped of their branches and stacked one on top of the other, with two pairs of rather shaky vertical poles holding them in position. It looked like one of the obstacles our uncles had their gamekeepers put up for our annual show-jumping contests. I have always found these competitions ridiculous. The sight of those great masses of horseflesh painfully hoisted aloft by cousins in pink coats and black riding hats, to the affected oohs and ahs of their womenfolk, has always struck me as parody. The only thing missing is a helmet, and a lance flaunting their ladies' silken favors.

Zigomar poked his long nose and shortsighted eyes over the barricade. With his rifle in the crook of his arm and his Confederate headgear pulled low over his eyes, he seemed to be hesitating between parts. But seeing that he was no longer the only one playing, he got into action. Thus are armies formed.

"Greetings!" he said, very grim — very gruff and Spartan.

"Greetings!" I echoed. "It's me. What's all this?"

We were both a head and shoulders taller than the barricade.

"Antitank obstacle!" he answered in ringing, soldierly tones.

We had never seen antitank defenses. They were a hundred times less familiar to us than an interplanetary rocket is to the young of today, who see such things live and in color on TV. We might have seen a blurred black-and-white photo in a dog-eared magazine, months out of date and relegated to the toilets — for it was there that the gently nurtured young of Touraine read that kind of magazine in those days. In any case, what we knew of such obstacles was not very different from the version Bertrand had built on Blue Island. The French army itself (on this point as on many others) was scarcely better informed than we were. To us that obstacle at the approaches to Blue Island looked formidable, impassable.

"Magnificent," I said, convinced. "And that?"

I was unfamiliar with the flag, homemade, which dangled for lack of wind from a hazel pole mounted on top of the barricade. In our games on Blue Island we generally flew the blue and red Confederate flag with its starred Saint Andrew's cross, which matched our caps and came from the same kit. This one was blue, white, and green, in horizontal stripes.

"One of Bertrand's ideas," said Zigomar. "You have to salute."

"What am I saluting?"

"I don't know."

I saluted the unknown flag. Since then an unreasoning passion, always tempered by the coolest detachment, has remained with me with regard to emblems of all kinds. I respect the French flag, that of my country, naturally. But I am a born waverer, not a patriot. I can see myself going willingly to my death, a song on my lips, enveloped in the folds of just about any flag as long as it looks good. But I must acknowledge that on the one occasion when that chance was offered to me — on this very day — I backed down . . .

"Where is Bertrand?" I asked.

"He's arming the reinforcements. He wanted you to join him as soon as you got here."

What we called our "reinforcements," I would remind readers, were our eight phantom soldiers, scarecrows in old greatcoats and First World War helmets, armed with old rusty firearms. They had already been guarding the northern "front" of the island for a week.

"And Pierrot?"

"On a secret mission," said Zigomar in his formal voice.

"Hidden somewhere along the main road. As soon as he catches sight of the Germans, he'll head back. We'll have some warning."

I felt a little twinge in my heart.

"Which doesn't mean they'll be coming here," I remarked to bolster my own morale.

"Quite true," Zigomar nodded.

From the glance we exchanged, I deduced that he too was counting on this unlikelihood.

Usually when I got to Blue Island after Zazanne arrived, she would come running to welcome me, her heavy breasts bouncing beneath her blouse. But not today.

"We haven't seen her," Zigomar explained, "and I don't think we'll see her at all today."

A new twinge in my heart, sharper this time. Almost of panic. Only now can I analyze it with any precision, for at the time it was just a confused feeling, an indefinable foreboding of dark things ahead. Staunch, solid Zazanne was not made for dramatic roles. Fate had taken this into account and had created no such roles for her. Her absence today seemed to herald trouble, and it was this I sensed, but at the same time I was not sorry she was absent. Maïté alone could have played the part of Woman that morning. If Zazanne had been there I would have wanted Maïté to send her back home . . . And I never knew the reason for her absence. Her father, most probably. It is no longer of any importance, and I have not laid eyes on Zazanne since.

I still had to ask about Maïté. I remember the sound of my voice, full of a trembling solemnity as it asked the simple question. Henceforth nothing involving Maïté would be simple.

"And Maïté?"

Zigomar seemed abruptly to have become mute. With a gesture as forced as the tone of my voice a second before, he pointed at the river, upstream from the bridge, that magically self-contained universe of greenery, sand, and water that Bertrand had appropriated at one swoop the previous summer. I leaned over the side of the bridge. I have already mentioned the curiously feminine nature of the Mulsanne at that spot, the near-hypnotic glimmer of the flowing water, which looked like long green strands of hair swirling along in the current. An oblique ray of sunlight passing over the treetops illuminated the whole river as far as the end of the island, and I saw Maïté's blonde hair mingling with the swirling strands of green. She was swimming slowly, eyes closed, as though performing some sort of ritual. Her movements were supple. A siren's movements that exposed, naturally and without the least ostentation, successive parts of her body: I realized that she was naked. I understood Zigomar's silence. As for me, I am utterly unable to describe what I felt, except that it was not anything I was used to.

Arriving at the westernmost point of the island, the boundary of our domain, she emerged from the water, standing on a bank of white sand, sparkling droplets of water cascading from her hair, and walked to the beach at the confluence of the two branches of the river. There she lay down on her back, her arms a little apart from her body, completely still. She hadn't said a word. I don't even know if she had seen me.

Shaking myself out of my stupor, I asked, "What's happening?"

"I don't know," said Zigomar, the very picture of perplexity. "When I turned up an hour ago they were both already here. They were kissing in the water, like in the movies. Then Bertrand came to join me and Pierrot got here. We discussed our battle plan. He's got some wild ideas, you'll see. I started to wonder about all this, but I said I'd be here and I'm keeping my promise. You saw her, she looks like a sleeping statue. Every so often she goes back into the water. She doesn't dive in the way she used to, she goes in slowly. She cups her hands and pours water on her forehead. As though she's baptizing herself. Then she swims for a few minutes, making no sound at all. She swims upstream as far as the bridge, then she comes back down here and lies down for a while without moving a finger . . ."

"And always . . . always undressed?"

My vocabulary left me few choices. "Stark naked," the usual cliché, was unworthy of Maïté and did not even come to mind. "In the nude" was no better, too light, not serious, with a vaudeville ring to it. Maïté was therefore neither "stark naked" nor "in the nude" but just naked, and that word would not do either: it was too charged with femininity to be easily spoken those millions of light years ago.

"Always," Zigomar replied soberly. "As though she's forgotten she ever wore anything but her hair. I don't understand it at all. Do you?"

I didn't, except that Maïté was now even further removed from us; she was silently moving along the outer frontiers of mystery. All of this meant something, naturally. Except that it has taken me forty-seven years to find out exactly what. When the time comes I will reveal it. I will tell how and through whom the key was given to me.

Zigomar's metaphysical preoccupations were over.

"Bertrand is waiting for you on the northern footpath."

"I'm off."

"First hide your bicycle, those are his orders."

I executed them.

"And take your weapon," Zigomar added, all spit and polish, as he handed me the Bosquette 5.5 he had taken from our old iron chest, which lay open at his feet behind the barricade.

There were boxes of cartridges in the chest, and bandages, cotton, the compass, a profusion of matches, chocolate, cookies, some string, a coil of rope, a map of the island and its "firebases" (which resembled the pirate's sketches in Stevenson's *Treasure Island*), all the paraphernalia of our old games, but also cartridges much bigger than ours, 12-gauge buckshot, old Kolb's ammunition. You could kill a wild boar stone dead at twenty paces with that. Even a man . . . I felt myself grow pale.

"This is war," Zigomar said to me with a friendly wink . . .

I found Bertrand sitting cross-legged on the ground at our second "firebase" in the undergrowth on the riverbank, a "covering" position for the bridge and a "staggered defense" for the shoreline, held for the moment by two of our "soldiers" standing stiffly in the reeds.

"Greetings!" he said to me. "Well, this is your sector. You'll be relieving me. You're late. Do you have an excuse?"

"I didn't wake up."

"It's true," he said, "you're a heavy sleeper."

This was the only time he alluded to my defection of the night before. I sensed that he didn't want to discuss it. He had wound a cartridge belt around his waist, loaded with shells as if for a partridge shoot. On his gray Southerner's cap he had pinned three silver captain's stripes that must have come from a uniform once worn by his father, Squadron Leader Carré of the foreign legion, exiled to the Chinese border, at the outer limits of the venerable Western world's death struggle. He did not look in the least ridiculous in all this accoutrement, just supremely at ease in his role, his eyes shining insolently, outwardly happy, a little arrogant perhaps, but who would hold it against him, considering the heights he was about to reach? Other instant promotions, at the Liberation four years later, would lack that natural elegance and were not bought at the same cost. I at once noticed Kolb's gun leaning against a tree, an antediluvian weapon with hunting scenes, boars, and antlers engraved on its stock, as well as a series of little notches burned in with a hot iron to commemorate the Prussians cut down in the Vosges forests in 1871. I later learned that Kolb was dying of old age at that precise hour and that Aunt Sophie Majorel, at La Jouvenière, sat perched on a stool from the stables, holding the dying man's hand as he babbled in delirium about "Master Bertrand."

"You know how to use it?" I asked.

On the bridge railing was one of our cans, a last survivor of our shooting exercises; it must once have contained peas. Bertrand stood up, raised the gun, took aim, and winced at the recoil. It made an impressive noise, putting flocks of birds to flight. The can disappeared, blown to pieces.

"You see, I've been training," he said gaily. "But don't count on my poor shooting to get you out of this."

Which seemed to me an ill omen, above all because he was smiling. I knew that smile from bitter experience. Then he returned to the work he was doing when I arrived. Armed with a schoolboy's ruler and a pocketknife, he was carefully measuring and cutting the tarred-paper fuses of a box full of fireworks. He had already prepared a good fifty fuses; they ranged from half an inch to four inches in length. The firecrackers were about the size of 8mm cartridges; packed in a crude blue box, they were the kind small farmers call "crow scarers," sold at seed merchants and at farmers' coöperatives. Designed to scare birds off fields at harvest time, they produce a terrifying din. We had already used them for practical jokes. I saw what he was planning to do with them.

"You think we'll stop the Germans with those?" I asked.

"Stopping them is too much to ask for. But they'll definitely be surprised. One of these makes as much racket as a soldier's gun. Imagine a whole barrage of them! I've set the fuses to go off at intervals from thirty seconds to three minutes. I did a trial run. If there are three of you it takes only twenty seconds to light them all."

"Who'll give the signal?"

"Maïté, from the keep. She'll see them coming from a distance."

It will be remembered that the de Réfort keep, an old stump of cracked stones standing just higher than the trees around it at the top of a small hill, gave a view of the whole road in both directions.

A question was worrying me.

"Will you be shooting at the same time? You could easily kill one of them if they aren't careful."

I no longer thought much of our chances.

"This is a battle, not an ambush," he said haughtily. "We warn them first. Then we fire."

"It'll surprise them for sure," I said, rather astonished myself. "And then?"

"That will depend on them. If they insist on coming through, we fight!"

For a moment I closed my eyes, appalled. I could already hear myself, arms held high, running toward the enemy: "Don't shoot! We were just playing . . ." At five foot five and with hairy legs, would I be believed? Would I even be understood? The Coward's Prayer rose to my lips . . .

Bertrand patted me gently on the shoulder. Looking at his watch, he said, "Your chances are getting better, chicken, maybe they won't come."

"What are you talking about!"

I must have looked like a plucked rooster desperately seeking to allay suspicion. Whatever my subsequent behavior, perhaps that last burst of pride redeemed a little bit of it. Had I been listening to myself at that moment I would have tucked my tail between my legs and fled.

We made the rounds of the "firebases." Bertrand laid his firecrackers in bundles at the feet of our stolid sentinels.

"Let's not forget the matches," he said.

We returned to the barricade where we found Zigomar and Maïté. She was wearing brief white shorts, a black short-sleeved blouse with its tails tied around her waist, sandals, and that was all. She extended her hand to me almost ceremoniously, without a word, keeping her distance. The heart has its rituals.

"Nothing from Pierrot?" Bertrand asked.

"Nothing," Zigomar answered, his military delivery masking his relief.

It was a little past ten. Zigomar and I looked at one another, not really knowing what to say—or rather knowing only too well, but increasingly intimidated by Bertrand. I am reluctant to fall into the trap of clichéd language and sentiment, but there was no getting away from it: Bertrand seemed to have risen to those heights where exceptional beings escape forever from the common ruck. He sat a little apart from us on a tree trunk, with Maïté by his side. Nowadays the almost blatantly exhibitionistic freedom of young couples is inescapable. Nowhere—not in the street, not in public places, not in recreation areas, not in living rooms—can we hope to remain in ignorance of the smallest detail of ephemeral unions proclaimed to the rooftops as if they were a new bill of universal rights. I do not allude to this in order to criticize it or to wish it away, but to stress a difference. Light-years, as I said . . . Well, Bertrand and Maïté were not holding hands, nor did they have their arms around each other's shoulders or necks or waists, their limbs were not intertwined, their arms or thighs were not touching, they were not even looking into one another's eyes, they did not intrude on the outside world with sweat, saliva, or ecstatic physical commonplaces; in fact there was no visible physical complicity between them at all. Yet they gave off almost palpable waves, carnal, sumptuous, enveloping them like a nuptial cloak. And on top of that they were beautiful. I have said that quite often enough . . .

A light wind stirred the folds of the flag, and in a moment it was fluttering as a flag should when great deeds are in the offing.

To fill the silence, I asked, "Why not the French flag?"

"Chabannais's flag? You're kidding!" said Bertrand.

"Or the South? It's got spirit. We're used to it."

"That was last year, when we were playing."

And we weren't playing anymore? What might that mean? I mechanically wiped my moist palms on my trousers.

"All right then, but what flag is this?"

Bertrand leaped to his feet and stood with his legs astride and his hands on his hips. With the last rueful smile I could muster, I thought, Here comes another speech . . . I was not mistaken.

"Bertrand Carré, only son of Squadron Leader Carré," delivered this curious declaration to Zigomar and me- "It is the emblem of Chance. We took what we could find among my aunt's old bits of cloth. There was no longer any red, or yellow, or black, or orange. Of course I could give you dozens of explanations, make up any number of symbols . . . Blue for the sky and for happiness, the awareness of a great beyond . . . White for purity of spirit and the mystery of a destiny not yet written, the blank page of the future, the virtues of oblivion . . . Green . . . Green for what? For mankind's hopes? For the trees and their vital forces or for the water in the stream? Nonsense! None of that. The flag has no meaning. It's mine."

"Hold the applause," Zigomar whispered, stunned.

Pierrot's whirlwind entrance spared us the need for further comment.

"I've seen them! I've seen them!" he shouted as he pedaled toward us.

Bertrand nodded to Maïté.

"Your turn now," he said.

She went off at a run along the trail leading to the keep. About a minute later we saw her blonde head above the undergrowth on the hill. It was then ten-fifteen.

As soon as he had caught his breath, Pierrot told his story.

"I was sitting on a wall just outside Petit-Bossay. I had felt them coming for a good long while. First, there was nobody in sight. Since morning I hadn't seen so much as a cat. Then there was a funny silence. I could almost hear myself breathe. Then every dog for miles around started barking at once and about twenty motorcycles with sidecars came into sight, moving slowly. The riders were in green, with their sleeves rolled up, wearing helmets and carrying guns slung across their backs. They didn't look too worried about anything. They said hello as they went by, and like a fool I answered. The last one was a bigwig, an officer, with a cross hanging around his neck and some kind of braid on his shoulders. This one was really on vacation, lying back taking it easy in the backseat of an open car. You should have seen that car! With a big radio antenna. He said hello too. Then came the tanks. I counted twenty. With those treads of theirs, they sound just like a harvester, only much louder and the engine at full. The men in them didn't look worried either. They were sitting on the edges of their turrets, getting a breath of fresh air. In black this time, with tight pants, short jackets, red epaulets, and cloth caps with visors, also black."

"Naturally they said hello," said Bertrand.

"Naturally."

"So they think they're in conquered territory. Go on."

"That's all. Except for two trucks in the rear, with about thirty men in helmets and green uniforms. There were others, but they were out on the highway. You could hear them for miles. I didn't want to check . . . Anyway, they all stopped in the square. They didn't seem to be in a hurry. They bought postcards at Magloire's. The mayor came out in his best blue, white, and red sash with the police, all buttoned up in their Sunday uniforms. Some of the men in the tanks laughed, but they got their heads chewed off. Then I left."

"Have they moved on?"

"I don't know. I don't think so. They turned off their engines. They wanted to buy beer and chocolate. The officer in the convertible unfolded a map. But if you want my opinion, let's hope they don't take the wrong fork at the crossroads, let's hope they keep going to La Roche-Posay. If not . . . There's no reason, anyway. The road to La Roche is paved and has signposts, and ours looks like a forest trail—but I'm still scared, Bertrand. Believe me, I'm scared!"

He wasn't the only one. Zigomar had turned pale. My mouth was dry.

"If you want I'll give you your freedom," said Bertrand. "Quick! Make up your mind."

He did not add anything unkind. His offer was sincere. He was releasing us from our allegiance to him. He had long ago made up his mind about us. But at the same time he had us trapped, with his grand, generous airs and his way of letting us know that in the end he would do quite well without us, that we didn't count . . . It was skillfully done. It seems to me that I have already alluded to that idiosyncrasy of Bertrand's: he had to have an audience—and he resented seeing it slip away before the final curtain.

Naturally, Pierrot showed off. "I'm afraid but I'm staying!" he announced bravely. "Me, too," said Zigomar.

I confined myself to a nod. I would have been incapable of uttering a sound.

"Good," said Bertrand as though he had never doubted us. "Now comes the waiting."

The old binoculars from the chest were his by right. He trained them on the keep, checking that all was well there.

"Can you see her?" we asked.

"Yes."

"What's she doing? Does she look as if she can see anything?"

"No, nothing. Still nothing."

The waiting lasted until eleven. I don't know what kinds of thoughts went through Pierrot and Zigomar's heads or how they managed to master their fear. We have never discussed it because I never saw them again. Another hour and they would be gone from my life forever. As for the weight of my own terror that morning, I have borne it all my life, a stifling sense of oppression, a paralysis of the will, a bleak hopelessness, every day at the same hour. I finally grew accustomed to it. It would dissipate within the hour. But when memory returned to me on my convalescent bed, my "eleven o'clock appointment," as I had baptized it, was canceled for good. That at least was an achievement. To explain it I would probably have to lie on a psychiatrist's couch, but I have nothing to do with such people: they would land me in a fight with myself . . .

Bertrand's voice broke the silence. With the binoculars glued to his eyes, he was questioning the fates.

"I think that this time — " he began.

"Don't say it," begged Pierrot.

A cry of triumph answered him.

"Yes! Maïté is coming back! She's already on the road! She's running! I knew it! I knew it! They're here!"

I recall a kind of transfiguration. I don't know any other word strong enough to describe his jubilation. In the far-off days when the human spirit triumphed unaided, this was how kingdoms were built, worlds conquered, and peoples brought to worship the presence of the divine . . . We came back to earth.

"Let's go, on the double!" said Bertrand. "Everyone to his post. Don't forget the matches! I'll light the first and you all light your own immediately after. Take cover in your firebases. Hug the ground. You must not be seen."

Maïté ran in long strides, in silence, the way she did everything. We didn't hear the sound of her feet, which seemed hardly to touch the ground. This was not fleeing but the trajectory of a blonde comet, its dazzling form intensely pure. Her face showed no signs of effort. She smiled. There was a narrow zigzag passage to let us squeeze through the barricade.

"Three tanks," she announced. "And two motorcycles. I think they've halted."

She disappeared behind Bertrand into the cover of the path we had cut around the island. Perhaps a minute passed; then the silence was broken by the clanking of treads described by Pierrot. They came closer, growing ever louder, as though grinding along inside my own head.

When destiny knocks to announce its arrival, one tends to look at one's watch: three minutes after eleven. By one hour, more or less, destiny had refused the reprieve I had asked for.

On this same day, June 21, let us return to Lieutenant von Pikkendorff's log as he advances along the right-of-way within sight of the keep, upright in his open turret and inhaling the forest air.

From this point in my story I will be using this log more and more. It is first of all significant that in its fully drafted form it comes to an abrupt halt on that day, June 21, 1940, and does not resume until January 1941, in Greece and then in Crete. From there on it is no longer anything but a succession of brief notes, hardly enough to trigger memory, recorded as if the writing was a chore and dealing only with Frantz's campaigns, combats, assignments, wounds, decorations, and leaves (which he usually either refused or spent in officers' rest centers). I have picked out the following: July 1941, Smolensk, Knight's Cross, wounded in the legs . . . September 1941, Kiev, promoted to the rank of lieutenant first class, wounded . . . February 1943, Yugoslavia, promoted captain, Iron Cross with oak leaves . . . November 1943, Rome. January 1944, Monte-Cassino. June 1944, Normandy . . . August 1944, Paris, where Major von Pikkendorff, missing an arm and, at twenty-four, holder of the Iron Cross with crossed swords, is assigned to the administrative staff of General von Choltitz, commander of the Paris region . . . No commentary. It is as if this man, who had taken willing, almost literary, pains to write well, reaping pleasure from his words and clearly eager to leave a living picture of himself, had suddenly stopped observing and speculating, had lost interest in the events of his own life. After reading the first part of his log, one has the impression that at some point he had suddenly ceased to live.

As for his entries for June 21, they too are different, and the reason for this becomes plain when one reads the entries that precede them. In several places the June 21 text has been altered. There are passages that have been rewritten several times. Lieutenant von Pikkendorff took pains to put his feelings into order. Occasionally he hesitates to express his innermost thoughts. There are deliberate omissions, with at least two pages torn out (the tear is visible along the inner spine of the notebook), as though he wanted to jettison ballast in order to master his own truth. I am not even certain that this entry was actually drafted on the day of June 21. More probably several days later, a time lapse that might have injected circumspection into his reminiscences. Nevertheless it is headed "Evening of June 21, La Roche-Posay." Finally, the facts that Frantz relates, most of which I witnessed and which I will in my turn interpret, occur within a span of barely forty-five minutes, from three minutes after eleven, as I have already said, to eleven forty-five (noted with military precision by Frantz). Such an eruption of events in so short a time, with storms breaking inside him, veils torn from unknown inner landscapes . . . Keep in mind that he was not yet twenty. He was dumbfounded by events, amazed. His facility with the pen reestablished some kind of balance, but we have to be able to read between the lines.

Here is how he began, still in French:

I still see those fateful crossroads. They lay nearly two miles past the village whose mayor — a sorry, ridiculous character — inspired such gales of laughter from my men as he insisted upon his peaceful intentions. He was in character from the word go. All that was missing was the key to the village, presented to us on a cushion. I truly have no luck with my French half . . .

Major Stockenfeld had halted the battalion. His command car was stopped right at the crossroads. Fate's dark switchman. We had been moving in short spurts since morning. When his officers asked why, Major Stockenfeld answered that there was now no better task for the glorious First Tank Battalion of the Third Westphalian Regiment of the Fifth Panzer Division, the most decorated of all Germany's armored units, than to listen to the birds sing. That's his sense of humor. This war is over. He was impatient for the next one. He was bored. He's thirty years old. In fact, the battalion's marching orders warned us not to cross the Creuse before the next day, June 22, and to establish our night quarters at La Roche-Posay.* Without an enemy facing us, we were like school kids playing truant. We had only begun these maneuvers to avoid stagnating. The sleepy little village of La Roche-Posay was to be the target of our exercise. Major Stockenfeld, his face buried in the map, is planning a pincer-assault, with diversionary attacks and encircling movements by light patrols. My platoon has been selected for one of these.

"Pikkendorff," said the major, "look at this trail . . ."

He pointed it out to me, a fairly wide dirt road, with embankments, under the trees, then returned to the map. "According to this it's an abandoned railway line. That's yours. You cross two small rivers, the Mulsanne and then the Claise at the village of Chaumussay. Advance to the village of Barrou and then to Chambon, just to impress the natives, and occupy Rouvray Château, which I believe is a charming little spot. It's on the banks of the Creuse, where I intend to spend the night and invite you to dinner. Be at La Roche-Posay by four o'clock. That's three times as long as you'll need. Go on, it'll get you out in the fresh air for a while . . ."

Young officers are always glad to be given free rein, to be allowed to become temporary kings of a brief space of earth and time, masters of the universe . . . We were happy as children, and why not? I was exactly nineteen when I left Würtzburg** with my first silver epaulet. I started there at seventeen. It was very likely at that time that I left

---

*Cf. footnote page 135.
**One of the officer-training schools of the German armored corps.

my childhood behind. And I never even noticed. There must not have been a perceptible difference. Only today is that transition apparent to me. And since then, commanding three battle tanks and fifteen soldiers, I have exercised my power to the full. I do not regret the use I have made of it . . .

We advanced slowly along the trail through thick undergrowth. I led the way. Kleist and Ludolf, my tank commanders, made irreverent use of the platoon's radio frequency. "We're out picking wild strawberries in fifteen-ton M3's," I heard Kleist guffaw, and Ludolf answered him, chuckling: "The good life!" I shut them up. They interrupted my daydreaming. Like all subordinates, they are incapable of daydreaming. They lack the capacity . . . We had raised our turret hatches. Turning around, I saw Ludolf (a bad-tempered man) stiffen in his turret as though on parade, probably angry with me, while Kleist combed the landscape through his binoculars. It was better to maintain a bit of discipline.

Not that there was much to see. We were out in the countryside without a living being other than pairs of pheasants taking to the air under our noses. Peace. Boredom. Meadows, copses, a little clear-flowing stream, low green hills . . . At a curve in the trail, a little above us, I saw an old ruined tower emerging from the undergrowth. Nothing about it to recall the Pikkendorff castle in Swabia, dark and sinister, brooding over a dreadful landscape, a place my mother hated, a place I had not thought about until this moment. From it, over the centuries, generations of raptors had spurred out to battle. My father, a diplomat, was the first Pikkendorff not to wear a uniform, and my mother . . .

Kleist hailed me. He had spotted something. His voice jarred my headphones. If he were speaking French, I thought, my ears would suffer less.

"Compass bearing forty-five, Lieutenant. Thirty degrees to your left. That heap of stones up there. There's someone there."

I trained my binoculars on the castle. Behind me I heard the creak of revolving gun turrets. Like the good fighting men they were, Kleist and Ludolf were turning to face the foe. They had already calculated the height. Their machine guns were ready to chatter.

"Wait for my order," I said.

I didn't see anything. That poor worn-down fort, heavy with past regrets, with only its pride to keep it standing until the Revolution and probably taken apart stone by stone since then by mean-spirited peasants, seemed innocent of danger.

"Lieutenant, a girl!"

I halted the platoon. I didn't like the greed in Kleist's voice. Did he believe that he and I would go half-shares on some tender prize of war?

"A girl!" Kleist repeated, and in his voice I heard long abstinence badly endured in a conquered country. "A girl, Lieutenant . . ."

A blonde head bursts out of the woods. There must have been a footpath leading down from there, for I suddenly see her emerge onto the trail thirty yards ahead of us. She turns back only once before speeding away. For one split-second she freezes in her tracks, standing sideways to us in a graceful posture that involves her whole body, poised like a statue on one arched foot, keeping her balance with her arms like a dancer, her face turned toward us. She is smiling. Impossible to guess for whom that enigmatic smile was intended, for she immediately begins to run, and we see no more of her but her long bare legs and the undulation of her hair.

"Not much covering her ass," Kleist remarks.

I checked my angry reaction. Even in French, his words would have gone down badly as a description of that young and beautiful girl, who seemed to me to have fallen from heaven. In German the words were almost filthy; they pained and disgusted me, but I must admit that Kleist was not wrong. I had seen more skin than clothing; nevertheless, his heavyhanded appreciation of the beauty we had glimpsed put a damper on my feelings.

The girl disappeared around a bend in the trail. She had been on a walk. She had not expected to find us there. She had been afraid. We would not see her again. I felt a great emptiness in my heart.

I seized my microphone.

"Move forward, slowly."

"What if it's a trap?" asked Werner, my gunner, who had watched the whole scene.

I shrugged, and dreamed of the beautiful stranger . . .

THE END IS in sight.

There will be gaps, blank spaces, in the scene I shall now attempt to describe. My memory had returned in force, no doubt about it, but was unequal to reconstructing this particular scene with the same sharpness as the ones that precede it. Moreover, this last intellectual reticence corresponded to a renewed onset of my illness, with high fever and delirium. They even had to tie me down at night, for I was fighting to get out of bed and flee. The only explanation I can offer is that a part of me had waged a fierce and painful battle to keep certain specific memories forever buried in the place where, consciously and unconsciously, I had long ago decided to hide them.

As I have said, it began with sound, that frightful clanking of treads, made even more unendurable by the fact that the tanks had not yet rounded the bend in the trail that hid them from us.

Lying flat in my makeshift "firebase" (I use the term, but I wasn't playing anymore, oh no! I wasn't playing . . .), dead still, paralyzed by fear, mesmerized by the stretch of gray trail gleaming through the reeds on the shore, I waited. I didn't even have the courage to recite the Coward's Prayer, that bleak shaft of private humor I sometimes aimed at myself. Instead I began to reel off terrified Hail Marys . . .

Then I saw them. First one, then two, then three, with two motorcyclists in the lead. Suddenly their gray-brown steel bulk filled the whole trail. More than the guns pointing like the fingers of God and the machine guns set flush with the hulls — they seemed oddly blind to me — it was the treads that terrified me. I could already see myself ground up under the weight of those mastodons, torn to shreds and reduced to a bleeding pulp by those linked fangs pursuing one another in a never-ending circle. My only consolation, and a feeble one at that, came from the black-uniformed men visible from the waist up as they stood in their open turrets. Perhaps they would see us before the tanks rolled over our bodies? Despite their sinister uniforms, they didn't look particularly bloodthirsty. Perhaps they would stop their machines in time? The man in the turret of the first tank wore silver epaulets. A young, almost childlike face, the eye watchful under his visor, but nothing like the gaze of a predator. I put all my hopes in him. If I stood up among the reeds, waved, shouted out my age and my name, anything to attract their attention, surely he would spare me . . .

But Bertrand was lighting his first firecracker, and I no longer had the time. The island rang with the explosions that reverberated from three of the "firebases." Zigomar and Pierrot were stoutly doing their bit. They had proved worthy. Amid an apocalyptic din (at least that was how I perceived it), I lay plastered to the ground, my face buried in the earth into which I longed to disappear. The air smelled of gunpowder and of smoldering wicks. Explosions went off all around me.

"Well, coward?" Bertrand shouted.

It was meant for me. Trembling, I struck a match. I had to make three attempts. All the same I managed to light all my bundles of fuses, and the series of explosions that followed seemed to sign my death warrant. That was my last act of courage.

It takes longer to describe than it took to live through it. It was all over very quickly. Nevertheless, our "sustained fire" kept going for a few moments thanks to Bertrand's calculations. I was flat on my face once more, watching despite myself, my eyes horrified by what they saw on the far shore. With hindsight I understand Bertrand's triumphant yell. It was in fact phenomenal. The two green-helmeted motorcyclists hastily turned tail, skidding round in a cloud of dust and fleeing for cover at the invisible end of the column. The same reaction from the soldiers in black: they dived into their holes, although their tanks continued to make their slow way forward. What followed is like a nightmare to me, like a living descent into hell. Suddenly, like a snake sniffing out its prey, I saw the machine gun on the lead tank swing around, and a second later it fired a series of angry bursts, a dry rapid-fire staccato that mingled with the last explosions of our firecrackers, instantly stripping them of their illusory power. Bullets whistled over my head. I heard a hail of cracks and thuds in the trees above, snapping sounds and other more muffled noises that resembled a fierce rainstorm in a forest. Leaves and neatly severed branches showered down. A sharper, more metallic impact startled me. Something struck my arm and rolled to within a few inches of my eyes. It was one of the Great War helmets our sentries were wearing. One of the rusty flintlocks followed, its stock torn away, then the greatcoat and its supporting pole were severed at chest height, and the poor theatrical display fell apart like a suddenly disjointed puppet. The helmet had a round hole with a jagged edge that had not been there previously. I had forgotten my gallant straw brother-in-arms. He lay dead by my side. I patted his greatcoat as if it might contain a body, then burst into tears of nervous exhaustion.

The firing stopped.

The lead tank had halted just across the bridge, the other two following a little behind and somewhat out of line. The officer with the silver epaulets resurfaced. Leaning carelessly against the hatch of his turret, he considered our barricade with a faint smile on his lips. Other soldiers in black emerged in their turn, heaving themselves up out of their holes and sitting on the rims of their turrets, legs dangling. One of them lit a cigarette. The two motorcyclists had reappeared as well and were standing next to their machines. We no longer scared the Germans. As Bertrand had planned, they had been "warned." Now we had to find out how they would take it.

I heard a rustling and soft footsteps along the footpath. It was Bertrand, his gun under his arm, followed by Maïté, Pierrot, and Zigomar.

"Second phase of the operation," he whispered to me as he passed by.

Second phase? What second phase? Get out of here forever, that would be more like it! I fought to collect my wits. It seemed to me that after this first engagement we were supposed to "regroup" in order to defend the barricade. Still trembling, crawling on my elbows, I dragged myself along.

"Where's your rifle?" Bertrand asked me.

I had left it on the footpath. Better to die of shame than go back for it. Bertrand didn't speak to me again. I no longer mattered. Pierrot and Zigomar were watching the enemy through crudely made loopholes in the barricade. Judging by their pale faces and clenched jaws they weren't feeling any braver than I was, but they at least were in control. That is the difference between courage and cowardice. Bertrand pointed his gun barrel through a loophole. A brief burst of fire rang out, passing just above our heads.

"Don't be a fool," Pierrot moaned. "Don't you think this is bad enough already? What if they start moving again, or use the big gun . . . Please, let's get out of here."

"Yes, let's get out of here," Zigomar repeated in a tiny voice. "Listen, Bertrand, we've really done all we can . . ."

The answer rang out.

"You maybe. Not me!"

I believe that at this moment he already had a specific plan, a new, clear picture of the climactic scene that had so far eluded him. But even then it would not work without the cooperation of the enemy. And this is exactly what happened; once again, Bertrand was lucky. This last word may shock sensitive souls, but I stand by it. What a pity, such a nice young man . . . Bertrand was *not* a nice young man.

From the lead tank came a voice. It spoke excellent French.

"This is Lieutenant von Pikkendorff, Third Westphalian armored regiment of the German army. I am duty-bound to give my men the order to attack. To salute your courage and your splendid resistance, I offer you a last chance to retreat with honor, with your weapons and your flag. If your commanding officer will come forward I will talk with him here. You have one minute to make your decision."

The tone was a bit theatrical, without the slightest hint of irony. The lieutenant was playing, but seriously. As his log will show, he had understood as soon as the fireworks started who was facing him and with what, and he was doing us the honor (I am quoting him) of joining in our game. Bertrand threw us a triumphant look. Pride, if justified, can be a form of happiness. And here it was indeed justified. A fanfare of heavenly trumpets from the four corners of the firmament could not have given him clearer proof than that single voice addressing him as an equal. But just as quickly, Bertrand's face darkened.

"What does this Pikkendorff think?" he said furiously.

A question with two meanings. Bertrand had not voiced the all-important question he had just asked himself: "What does he think? That I'm playing?"

There is no other explanation.

"If I were you," said Pierrot, "I'd grab the chance anyway."

"Would you?" said Bertrand icily. "I'll answer him all right!"

"I'm coming with you," said Maïté.

They walked to the middle of the bridge where the German lieutenant was standing, legs apart, hands behind his back. Through a gap between two ill-fitting trunks in the barricade I saw only their backs. The lieutenant was facing us. Their voices reached us clearly from thirty feet. The German saluted.

"I am Frantz von Pikkendorff, lieutenant."

"And I am Bertrand, Bertrand Carré."

I waited for the speech: "I, Bertrand Carré, son of Squadron Leader Carré of the foreign legion and of Laïcha, Oriental dancer," etc. It didn't come. Yet this was the right moment for it. It was part of the game. It had been perfected, dressed up, polished, for a situation exactly like this one.

Zigomar gave me a worried look: "It doesn't look good. He doesn't know what he's doing anymore . . ."

But unfortunately, he did. He had deliberately forsworn our shared ritual. He was cutting the lines that bound him to us. He was leaving us.

Noticing the three silver stripes pinned to Bertrand's Confederate cap, Lieutenant von Pikkendorff went on ceremoniously and perhaps a little bit too formally: "A captain, I see . . ."

"Captain? At fifteen? You think so?" Bertrand answered as though mocking himself.

He had added four months to his age. I have tried to understand why. Fifteen is a milestone everyone recognizes. At that age you aren't a child anymore. Bertrand wanted this to be clearly understood. He spurned any indulgence or charity his tender years might inspire.

"That is of no importance now," he added. "For that matter, neither is the cap."

With a quick movement he threw it in the river where we saw it float beneath the bridge and disappear in the current. It seemed to me that the lieutenant hesitated. Perhaps, realizing that he was playing the wrong part, he was wondering why.

"As you like. And this flag?"

No speech there either.

"It's mine," Bertrand answered briefly. "You are standing on my frontier. You will not cross it."

He was tired of big words, as if he had wasted enough time already.

The lieutenant had recovered his composure. His voice was warm.

"Listen to me, both of you. I feel great friendship for you. I am happy to have encountered you on my path. Now go home, with your three comrades. You wanted to play with the German army and I think I understand why. The German army thanks you for it. I will remember you for years to come. You too, Mademoiselle . . ."

It might have been a compliment, an elegant and easy way of halting matters, of putting an end to the interview. No answer was needed. He bowed slightly as if about to take his leave, as German military etiquette required. For him, the affair was settled. So he was taken aback when he heard Maïté's voice.

"I am Maïté de Réfort," she said with the casual insolence and apparent absence of emotion that were typical of her. "I too belong to Bertrand Carré."

It seemed to me that the "too" had been pronounced as if detached from the rest of the sentence, after an imperceptible pause and slightly louder, almost provocatively. At the time, of course, none of this made any sense to me. Why throw her relationship with Bertrand at this German's head? What did it have to do with him? An actress to the end, Maïté was . . . In any case I was much too busy concentrating my last energies on the effort of not taking to my heels, for all this was happening at terrifying speed . . .

"Interesting," the lieutenant commented. Then, looking at his watch: "You have two minutes to leave the area."

The tank drivers had started their engines up again. The din filled the forest. The lieutenant had climbed nimbly back into his turret and now stood upright, leaning on his forearms, as though on the bridge of a ship. I measured the tank's height with my eye, then studied our ridiculous barricade, which suddenly looked as if it had been put together with matchsticks.

"Now's our chance to get out of here!" said Zigomar. He shouted: "Bertrand! Maïté!" Then to Pierrot and me: "What are they doing?"

For they had not moved. Now they were holding hands, but that only lasted for a moment.

Bertrand turned and called to us, "Quick, my gun!"

"He's crazy!" Pierrot moaned.

I cannot explain what I did next. It resembled the behavior of a panic-stricken rabbit whose instincts are turned upside down so that it heads straight for the thing threatening it. Even today it is wrenching to see myself as I was. Grabbing Bertrand's gun without a second's thought, just because he had asked for it, I ran onto the bridge, out of my mind, and tossed the weapon to him as I ran by. Then, raising my arms and shouting, "I surrender! I surrender! Don't shoot! I'm fourteen . . . ," I raced along the line of tanks like a whipped dog in flight, pursued by roars of laughter from the soldiers. But what crucified me was Maïté's look of disgust, which I felt boring into my back. Disgust. Not even contempt . . .

I collapsed at the trailside, convulsed by trembling and sobs, body and soul ripped out of joint. Then I heard a tremendous report. It was Kolb's blunderbuss. I saw Lieutenant von Pikkendorff raise his hand to his shoulder where blood was flowing. Then, with features frozen halfway between a smile and a look of boredom, he drew his revolver and fired. In the next instant the little sanity that remained to me vanished.

I have said that Frantz's log was crucial to an understanding of events. Here is how he described what happened:

Kuntz, my driver, saw it at the same moment I did.
"Straight ahead, two hundred yards!"
We had just rounded a bend, and the trail ran in a straight line to a bridge. It was blocked by something my binoculars suggested was more like an obstacle in a show-jumping contest than an antitank barrier. Not even amateur work. I wonder who thought of making a stand against German tanks here — and why, seeing that logically we should never have come this way at all. A flag flies over the barrier, not a French flag or the flag of any known country. I'm beginning to understand, I think. Over the intercom I warn my men not to fire without my order. I hear an exclamation through my headphones. This time it's Werner, my gunner. Nothing escapes his sights.

"Guess what we're up against, Lieutenant! To the right of the bridge, among the reeds. An army of scarecrows!"

I train my binoculars on the spot. Old helmets stuck on sticks! I didn't yet fully understand, but I was impressed. Thumbing their noses at the German army on this June 21, 1940, with the armistice just around the corner . . . At ground level, in the bushes on the other side of the water, I see figures moving, and suddenly the air is full of detonations. I dive into my tank. Instinct. Kleist and Ludolf do the same. As for my two motorcyclists, their panic is a pleasure to behold. We'll be laughing about it for months. I shout into the microphone, "Hold your fire!" I leave the hatch open. Outside the fireworks continue. "Don't you notice anything?" I say to my men. "Listen."

"He's right," says Werner. "No bullets."

Normally, they ping like bells on the armor plating. With such sustained fire, we should be hearing chimes.

"Firecrackers!" I say. "We're being attacked by kids with firecrackers! Wonderful!"

"Wonderful?" Kuntz asks, puzzled, as romantic as an ox.

I allow enthusiasm to carry me away. I have been hoping for something like this, praying for it with all my heart. A gallant last stand in the best French style. I didn't want the war to end before equilibrium was reestablished between my two halves. But I had imagined nothing like this. This is even better and more beautiful than I had hoped.

"A last stand, Kuntz! That doesn't make you happy?"

"Kids playing? It's absurd. We're wasting time. I'm past the age."

"I am not!"

I had assumed my commander's voice.

"Werner," I say to my gunner, "saturate those bushes for me. Elevation six feet. There isn't much risk. From what I've seen of them, that should be well above their heads. Try to bring down the scarecrows. They'll be very proud of this one day, these kids."

"Sir!" my second-in-command, Kuntz, protests.

I cut him off.

"You unimaginative sausage-eater! These French kids are paying us

the honor of playing with us. That honor will not be refused! Go
ahead, Werner."

Yet I was a bit worried as I watched through my periscope. If I were
to wound or kill one of these kids, I would have a hard time settling
it with my conscience. And then there's the blonde hair . . . After
twenty seconds I give the order to cease fire. In his cockpit, Kuntz
shrugs almost imperceptibly.

"And now, Lieutenant," he asks stubbornly, "do we flatten their
little barrier or do we play another game of marbles?"

"Halt just this side of the bridge," I tell him.

Kuntz's remarks left me ill at ease. Nothing to do with him. He's a
dolt. Incapable of abstractions. But for some time now, since giving
the order to fire, I had been telling myself that playing at war was the
same as mocking it or parodying it without glory — unless you
adopted all its rules, including the risk of losing your life and its cor-
ollary, killing. And these kids were simply playing — or were they? A
strange thought was going through my head, perhaps a legacy of the
Swabian Pikkendorffs, and I was not sure I welcomed it. I tried to bar
its way to the doors of my soul. I drove its will-o'-the wisp glimmer
back into the half-light of unacknowledged madnesses.

"Kleist, Ludolf, don't interfere!" I order my tank commanders.

Then I jump from my turret. I can see them behind their little
barrier. They look about fourteen. There are five of them, including
the blonde hair. I think back to my warrior ancestors . . . Whereas
now — no girls, no loot, what a dreary war! At least let's put an end to
this particular war, apparently winding down at this laughable bridge
where I've immobilized three of the immortal Fifth Panzers' tanks. I
was hoping I would rediscover my French side here. But I feel more
German than ever. One of the boys is pointing an old gun through a
slit in the barrier, an outlandish blunderbuss aimed straight at me.
He's pushing things a bit. Five seconds of Werner's machine gun at an
elevation of six feet to let them know recess is over. Too late, kid, no
more playtime . . . I deliver a brief and heartfelt speech to these
impudent cherubs, in French, with appropriate congratulations in the
best d'Artagnan style, following it up with an order to move out and
make room for the grown-ups. I add that we can parley if they wish.

At least I'll see what they look like up close. And what *she* looks like, of course . . .

That's what I wanted. That's what I got.

The boy who comes forward is almost as tall as I am. A bracelet gleams on his wrist, a sign of arrogance. Whenever I have looked at myself in a mirror, or in a girl's eyes, I've always thought I looked a handsome devil, attractive, with the advantage born of knowing it and of using it intelligently. But no mirror has ever thrown back at me so perfect a reflection as the one I see in this hotspur who doesn't even return my salute and who introduces himself as Bertrand Carré. He doesn't look as if he's playing at all, yet he's mocking me. I address him as though he were a baby, as though his cap, his stripes, his flag, had created a bond between us. I must have seemed false and ridiculous. I make declarations of admiration and friendship, of the "what you have done here is gallant indeed" variety — and he answers me as a man, telling me tersely and bluntly to leave. Thank God no one in my platoon understands much French. As for the girl — or rather the (very) young woman, for it's a woman facing me, unabashedly asserting her womanhood — her eyes never leave me. Gray eyes, intensely cold, like a shimmering ice floe, so empty of expression they are almost translucent, but masking a spark of impatience, a powerful buried desire to know. If I didn't know what happened next, nor how extravagant was the thought that came to me at that moment, I would swear she had come to see me die.

Our conversation could not go on. I had talked enough nonsense. Quite enough. They were both far too good-looking, let them go to the devil . . . I gave them two minutes to clear out of the area.

The boy merely shouted, "Quick, my gun!" For a moment after that they held hands, and that I didn't like, not at all . . . Then a great rabbit of a boy burst out from behind their "sweet little barrier," a gangling youth who came running across the bridge in clumsy bounds, looking panic-stricken. He had a weapon in his hands. A respectable-looking old gun that might well work.

I alerted Werner: "No shooting! We do not slaughter children with machine guns . . ."

Without breaking stride the rabbit tossed his weapon to the other boy (who caught it in the air), then fled down the trail past the tanks, howling comically, "I surrender! I surrender!"

It made me laugh. I was wrong. The blunderbuss gave an impressive and unexpected bang, as unexpected as the pain I suddenly felt in my left shoulder. I put my hand to it. Blood was flowing. The surface of the flesh was torn. Buckshot, probably. I looked at the boy with the bracelet, the one who had shot me, Bertrand Carré. He was nimbly reloading. I shouted to Werner, "Leave him to me!" And I didn't miss. With my revolver. Right through the heart. A small, clean, round hole. A nice shot, too, at that distance . . .

I killed Bertrand Carré. I see no dishonor in that. I killed him in combat. Of course there would have been many ways of sparing his life. I even considered doing so as I drew my revolver. What decided me to fire were the eyes of the young woman named Maïté de Réfort. There were three characters in the drama. Even if indistinct areas still remain, it was consistent with the dignity of all three.

I had Kuntz put a quick dressing on my wound and went over to examine the youthful corpse. We carried him to the side of the road and set him down on the grass. It was I who closed his eyes. The other kids stayed hidden. The rabbit must have been trembling like a leaf in a ditch somewhere. I sent Kleist to take down the flag, and we covered the body with it. Mechanically, I asked the girl what the emblem represented.

"Chance."

Her face said absolutely nothing. She appeared to possess the remarkable gift of creating a void within herself.

"You will tell his family?"

"I'll take care of it," she answered expressionlessly.

My wound was hurting. I was agitated, overexcited, swept by emotional confusion, and the blood was pounding in my temples. I looked at her bare legs, at the white skin touched with gold, at the slow, insolent rise and fall of her breathing beneath her blouse. I was unable to check the words that rose irresistibly to my lips.

"In another age, Mademoiselle, I would have claimed you as a spoil of war."

Her gaze did not waver. She went on looking at me with that icy, maddening curiosity . . .

Frantz von Pikkendorff's journal entry for June 21, 1940, ends there. The passages concerning Bertrand's death and Frantz's meeting with Maïté have no corrections. They have obviously been transcribed from a rough draft. The reader will also notice the frequent changes of tense. The past tense has been substituted for the present tense employed almost everywhere else in the journal. And it is from this section that the pages were torn.

I have already mentioned that the journal does not begin again until January 1941, when German forces moved into Greece. With one exception, however. A date and three words in the center of a blank page:

June 29, 1940. It is done.

It remains for me to add a quick account of what became of the "panic-stricken rabbit," that is to say of my miserable self, that day.

Apparently I made my way back to La Celle on foot, forgetting my bicycle, forgetting even that I had ever had one, running the three miles without stopping, guided solely by the instinct that leads frightened animals unerringly back to their lairs. I was unable to answer questions, unable to think. I jumped at the slightest sound, I trembled without apparent cause; and the rest of the time I stayed in bed, unspeaking, eyes dull. For a time people even believed I had lost the faculty of speech. I refused food. My aunt's doctor, who was immediately sent for, talked about emotional shock, the term used in country towns in those days for what is now called "depressive syndrome." It was a violent case. Beneath those bedcovers I rebuilt my entire life, and I had no intention of coming out from under them. Alerted by Aunt Melly, my mother (my father had been detained in Bordeaux by the offer of the position of secretary of state, which he ultimately refused) came to pick me up in a car on the day after the armistice was signed, June 26. She provided the details above.

I recovered slowly but steadily, until the day when my parents, returning to the events of that day, tried to ask me some questions. There was an immediate relapse. They didn't try again. Occasionally a memory would resurface, of course, but as soon as anyone attempted to grasp it, I would shut up like a clam. They even stopped sending me to Touraine for my vacations, even though milk, honey, and all sorts of good things unavailable in Paris still flowed there. I would have let myself be killed rather than go back. Giving up the attempt, my father bought a small vacation home near Vernon, in the Eure region, out in the countryside, where my mother spent many a boring day while I developed fruitful relations with the neighboring peasants: it meant that we would all eat our fill for the rest of the war.

And time went by.

I SAW FRANTZ von Pikkendorff again in Paris four years later, on August 23, 1944, to be precise. It cannot be called chance. He had an appointment with death, and I with my past and with Bertrand Carré's ghost, still lurking in the depths of my memory.

As everyone knows, the fighting to liberate Paris began on August 19 with an insurrection by a section of the Paris police, the first to raise the French flag in the city. The FFI (French Forces of the Interior) followed. The morale of the German forces under General von Choltitz was low. Few in number, they had pulled back into the various major buildings they had been occupying—the Navy Ministry, the Crillon Hotel, the Meurice, the Majestic, the Lutétia, the Kommandantur on the Place de l'Opéra—their determination dependent on the caliber of individual commanders and on the number and combat-readiness of the forces at their disposal. On August 21 barricades began to go up all over Paris, some of them undeniably quaint and nestling in romantic sites, with the intention of hampering the movements and communications of the Germans, who had in any case lost most of their will to move. The sound of fighting was surprisingly unobtrusive. As far as I can recall, that was the general picture.

I spent a great deal of time walking through Paris during those historic days. I had just turned eighteen. My school had patriotically closed its doors, and my fellow pupils had gone off to urge on the fighters from the tops of unsteady scaffoldings around the Latin Quarter. It will be understood that my memories of Blue Island paralyzed in advance any desire I might have had to join in the fight. What still surprises me is that the "coward" I had never ceased to be felt no fear. I was rather proud of that, discounting the fact that the whole of Paris was also out in the street, listening with a delicious thrill for the rare sound of bullets overhead. Most often nothing happened. A sort of *Threepenny Opera* with a patriotic theme. It has all been chronicled; there is no point in going over it again. It is part of the capital's Liberation routine, combining tales of heroism with tales of an equal number of savage and contemptible excesses.

On August 23 I found myself in the Tuileries Gardens by the Place de la Concorde. A glorious summer morning. Chance . . . Small boys were sailing boats in the big fountain. Young women sat on benches watching their children play. There were even nannies pushing ornate baby carriages and tottering old men watching the little boats. Farther away, near the open gate on the Place de la Concorde, a small troop of Parisian warriors in arms were crouching in the shelter of the stone parapet. They were shouting at the passersby, "Get down! Get down!" but no one was getting down. It was hilarious to watch those young volunteers, defending the motherland from a prone firing position among children playing with hoops, young women taking the sun, and old men chatting on benches and peaceably discussing the action going on nearby. I describe it as I saw it. Except for the somber glittering example of the Commune of 1871, Paris has never been Warsaw.

I approached the combatants and eventually "got down" myself in order to put myself at their service. They were excited.

They pointed at the German barricade on the rue Royale. From afar it looked deserted. I mentioned this to them.

"The bastards have taken cover!" their leader answered.

"I'll go look."

"You're crazy! And screw you!"

I wasn't playing war. Those who have read me this far will know that I had drained the illusions of that game to the dregs.

I wandered to the sandbag parapet blocking the rue Royale between the Hotel Crillon and the Navy Ministry where the swastika still flew. Other passersby crossed through the zigzag passage in order to get to the church of the Madeleine; they were checked at the crossing and had their pockets searched by a German officer. For it was an officer. I saw his silver epaulets with gold braid, his gloves—his glove, for he was missing an arm—and his collar with a cross, which gleamed like a gem, on a red and black ribbon. Two hand grenades hung from his belt. His boots were impeccably polished. He was alone. Absolutely alone.

"Your identity papers, please," he said to me in excellent French.

His face was extremely thin, his eyes set deep in their sockets. One of the sleeves of his tunic hung limp and empty at his side. I recognized him by his voice. He recognized me by my face, which was no cause for pride.

"The rabbit . . . ," he said in wonder.

That unflattering picture of me had therefore survived four years of hard campaigning on every front in the war. And I had tried so hard to erase it from my own memory . . .

"Have you any news of Mlle de Réfort?"

I had none. I didn't want any. I hadn't gone back to Touraine and she didn't live in Paris. I told him so.

He made no comment. Searching through an inner pocket of his tunic, he handed me a small notebook with dog-eared pages, missing its cover. It was his journal. Why was he carrying it on his person? Why did he give it to me?

"I will never again see Germany," he told me. "I had thought of burning it. But since you are here . . ." He looked at me with friendship and added only: "Pass."

I brought away from him a sad smile, which briefly lit up that ravaged twenty-four-year-old face, and the image of his single gloved hand attempting without success to rebutton the two top buttons of his tunic. I was shattered.

I stopped walking through Paris, and, in any case, the German garrison's capitulation was signed two days later, on August 25, 1944. Despite General Leclerc's precautions, a number of old scores were settled in Paris in those days; and there were a few savage reprisals against German officers who were prisoners of war — not all of whom, in fact not many of whom, were Nazis. The father of one of my schoolfriends served on Leclerc's administrative staff. He made inquiries. The answer, alas, was not long in the finding. During a transfer of prisoners, in the midst of popular agitation and fighting between rival French factions, Major Frantz von Pikkendorff was intercepted by a horde of eleventh-hour irregulars, slapped, degraded, spat upon, clawed until the blood flowed, beaten black and blue, and finally finished off with a bullet in the back of the neck.

I didn't open the journal. All I wanted, and as quickly as possible, was to bury myself again. I hid it in the back of an armoire, in a spot where I knew I would never again see it. When I left the apartment for the last time, to be married, my mother made me a gift of that armoire, along with a few other pieces of family furniture. Forty-three years later I had no trouble finding the journal again, going straight to the right shelf and removing the board behind which it was hidden.

In this I saw a sign of the hand of fate, a sign most writers seek in vain before buckling down to work.

FINDING THE JOURNAL made me want to get in touch with Maïté.

She hasn't gone by the name Maïté de Réfort for at least forty years. She changed both her first name and her last; but, as with Arletty or Chanel, French custom and Maïté's own extraordinary talent dictate a respectful "Miss" before her name, which thus becomes an integral part of her stage name. Although she told me herself that she didn't care whether or not she was revealed in the character of Maïté de Réfort—her "greatest role," she had added—I shall withhold this name. It is that of one of our most celebrated stage actresses (the postwar period gave us a dozen or so) and has graced the best theaters in Paris without a break since 1947. Montherlant, Anouilh, Claudel, Pirandello—she has served the greatest and has even redeemed other playwrights less worthy of her. People came to hear her voice, to see her eyes, her wonderful features. That gift which had struck Pikkendorff so deeply, "the remarkable gift of creating a void within herself," brought every role within her reach, but with that emotional remoteness, that silent intensity that was the mark of her genius. Yet she was also the first actress of her stature to appear completely nude on stage, in a Garcia Lorca play in the early fifties. She had arrived at this decision alone. There were strong repercussions, but no one cried scandal. People spoke of a "carnal liturgy," of the "participation of the whole audience in a kind of divine rite during which time seemed to have stood still. . . ." She never did it again, but those who witnessed that "sublimation of the flesh" — the very expression that was used — are unlikely to forget those minutes when they forgot to breathe.

I was one of them.

But I had closed my eyes. I was back in the clearing not far from La Celle, with Maïté standing, naked, on the wooden plinth of her own statue and Bertrand telling me, "A good reason to die . . ." She was replaying that same scene.

Of course I avoided seeking her out. There was in any case no chance that we would ever meet. She lived like a recluse, always in a hotel, and she changed hotels often in order to cover her private tracks. She refused interviews and photographs (except for stage photographs of the roles she created, under conditions she dictated) and never received anyone. She did not attend parties, fashionable restaurants, festivals, television studios, or fundraisers. She refused autographs and never replied to fan mail, handling her own affairs without the help of an accountant. So complete was the blank surrounding her private life that she became the most mysterious figure in Paris and the object of gossip and rumors both numerous and improbable which she never troubled to deny. As for me (and taking account of her reclusiveness), my small reputation made it unlikely that we would ever come face to face. I have always loved the theater with a passion, but I avoided seeing the new plays she starred in. I was no more anxious to reopen old childhood wounds than I was to seek out the hiding-place where Pikkendorff's journal slumbered.

Of course, over a span of forty years there had been exceptions. A revival of *The Dead Queen,* for example, and another of *Leocadia.*

Friends dragged me along despite myself. At the opening night of *Leocadia* — and this was just a short time ago — I was in the fifth row, dead center. Over the years my reputation had grown somewhat. After the Prince bowed out at the tenth curtain call, Maïté stepped forward to take her bow alone, as was her invariable custom. Running her famous gray eyes over the house like a sovereign reviewing her subjects, her gaze met mine and lingered there for a long moment. Everything told me she had recognized me. I didn't lower my eyes.

For me, it was a sign.

The very next day I left my card at the ticket office.

At six that same evening Maïté called me. The telephone accentuated the studied remoteness of her voice. She didn't give her name.

"Come right away," she said briefly. "To the theater."

She hung up. I noticed she had employed the same informal *tu* we had used long ago. I also noticed that she had not for a second imagined that I might have a previous engagement. She was right: there are some meetings you will sacrifice everything for.

Her dresser led me up a narrow little stairway hung with maroon velvet. It was newly redone, as was much of the rest of the building. At sixty-two, the lady was still making theater managers' fortunes. She received me in her dressing room, which was furnished like a drawing room. I recognized the Louis Quinze love seat, two armchairs of the same period, and a magnificent pair of silver candelabra from Aunt Octavie's boudoir at La Guichardière. On an English maple trolley were champagne in an ice bucket, a dish of caviar, toast, a basket of fruit. She nibbled occasionally. No makeup. A few wrinkles at the corners of her eyes and mouth told me she had spurned the evanescent reprieve of a face-lift. She did not get up when I arrived, but remained half-reclining on her chaise longue. She made none of the remarks that might be expected in such an encounter — two childhood friends meeting again after a lifetime's separation.

"Sit down. Would you like something to drink?"

I had loved that voice passionately.

"I've been reading you for years," she went on.

She mentioned the names of several of my books, accompanying them with terse, penetrating analyses and occasional praise.

"I find you in them," she said. "You haven't changed. You express yourself better all the time, your imagination is sharper, but it's no good putting up that smokescreen. I can see behind it: you're still afraid of girls."

She used exactly the same words as she had on Blue Island, when she was acting out the part of "runner" on the footpath by the riverbank. Did I really still give the same impression? In her presence, I have no doubt that I did. I kept quiet, happy to bask in the icy flame of those eyes.

I had loved those eyes passionately.

She gave me no quarter.

"You're still afraid of everything, of course, of life, even of yourself. That's why you celebrate noble sentiments, lost causes, sublimation. You amuse me . . ."

It was not pleasant to hear, but it was said without the disdain she had shown me so long ago. It felt like a resurrection.

A photograph in a plain wooden frame sat on a side table. It was the only one in the room. Indeed the first thing you noticed here was the absence of the trophies — posters, photos, models of sets, sketches of costumes — with which actors surround themselves. I couldn't take my eyes off it.

"Oh, don't be bashful," she said, "it's really me."

Maïté stood on tiptoe facing the camera on the little beach at the northern tip of Blue Island, her long blonde hair tumbling around her, her eyes closed, her arms upraised, her hands outstretched.

I had loved that body passionately.

But not one of us had owned a camera.

"Who took this picture?" I asked.

"Frantz."

I remembered the laconic note in the journal: "It is done."

"The twenty-ninth of June, 1940," I said.

"How do you know?"

I told her about the campaign diary. I had it with me. I offered to leave it with her.

"Why?" she said. "Keep it. Hand me the picture, would you?"

I rose and took it to her. She held it in front of her eyes, on her knees, like a book, or perhaps a mirror.

"He came back on the twenty-ninth of June," she went on. "Alone. On a motorcycle. Shining, beautiful as a god, German. Conqueror . . . The law of the conqueror. He came to La Guichardière and asked to speak to me. My father came to get me in my room. He was dumbfounded. I went down the stairs. Neither Frantz nor I said a word. I got on the pillion seat and he rode off like a meteor to Blue Island, passing through the village on the way. At sundown he brought me back. It created a frightful scandal. But I had no trouble shutting my father up. I just refreshed his memory of the scene at the barricade in the village, with the unfortunate captain and that ridiculous little gun of his that ended up in the pond . . . That same evening I was bitten by the theater bug and found out that I had a talent for it. I never saw Frantz again."

"But the photo?"

"It was sent from Germany a month later, without any message, without a return address. Besides, I wouldn't have answered the letter."

Something intrigued me.

"That picture of you. And none of him?"

"Naturally. Put it back in its place, please."

Hiding what I knew, I said, "Do you know what became of him?"

"I have never even asked myself the question," she said.

Not the slightest emotion in her voice, *of course*. I didn't insist. I wanted to get to Bertrand. So far we hadn't even mentioned him. I wasn't sure how to go about it. She looked at me searchingly, and I felt naked under her gaze. A hint of mockery appeared in the gray eyes.

"I see I'm shocking you horribly. Your fine feelings are ruffled. Because I could go from one to the other like that. Without being bothered by it! With Bertrand barely in his grave . . . But I would have done it on the very same day if Frantz had asked me to . . . My poor sweet simpleton, did you think it was single combat between two noble Christian knights? There was nothing of the Christian knight in Bertrand. Or in Frantz either, please believe me. They realized that at once, with the first glance they exchanged . . ."

I fought to restore order in my memories.

"Do you remember our games," she went on, "in the old bus at La Jouvenière? Bertrand always killed the enemy leader. It was an understanding between us. Unless a dead chieftain was laid at my feet there would be no Maïté for him. Only symbolically, that year. The next year, on Blue Island, we renewed the pact. We swore in the name of the river and its waters. We said our pact was sacred. The usual children's mumbo jumbo . . . On the morning of the last day, I decided to reverse the order of our agreement. I didn't try to hide it from you. It seems to me that I was quite clear. There was only one thing left for Bertrand to do — kill the first person to appear . . ."

And of course Mlle Maïté de Réfort had given herself the starring role . . .

The very first tear of her long career as an actress welled at the corner of her eye, hesitated before the enormity of giving expression to a feeling, then made up its mind and trickled slowly down her cheek. Maïté wiped it with a finger and studied it in surprise.

"You know what we were, the two of us, Bertrand and I?" she said. "The most unbelievable little monsters."

Her dresser came in. The hour of her metamorphosis was at hand. It took the grande dame two hours to get into the duchess's skin. I took my leave.

"We won't meet again," she said. "I don't wish it. What will you do with all this? A book?"

I answered that this was in fact my intention, if she gave her consent.

She swept the air with the back of her hand.

"Tell your story, my poor simpleton . . ." As I left she added one parting shot: "Don't forget to include Melly as supporting actress!"

I have never really understood what she meant by that . . . In any case, my aunt Melly Lavallée had been dead for three years. She was the last of my uncles and aunts to go to the grave.

I wanted to see the place again, quickly, without opening my heart. La Guichardière had been sold, by Maïté herself, her parents' sole heir. La Jouvenière as well, by Aunt Sophie. Introducing myself as a collector of antique carriages, I asked whether they might have an old coach or barouche on the property that they wouldn't mind getting rid of. I was told that when they had bought the house it had been empty and uninhabited. I avoided La Celle, hideously restored, where I might have run into Cousin Louis Lavallée, wealthy senator for Indre-et-Loire. Beausoleil and La Cornetterie were shuttered. No one spent weekends there anymore, and Zigomar Durand, pharmacist, was in the final stages of making his fortune in Orléans. As for Pierrot's farm, it had been spruced up, festooned with windowboxes, and decorated with white-painted wagon wheels and a fake wishing well: it reeked of a Parisian's weekend home at a hundred paces. Any goose liver consumed there would come from a supermarket.

Blue Island was the big surprise. In truth, very little had changed there, except that it was littered with soiled paper napkins and empty cans and bottles, the leavings of Sunday picnickers. Beside the trail just before the first iron bridge I stumbled upon an ordinary-looking gravestone, drab in appearance but inoffensive, like countless other gravestones the French erect in their pursuit of posthumous dignity. The date of its erection, carved on the stone, corresponded to Louis Lavallée's political debut, 1956. He had loathed Bertrand, who felt the same way toward him, yet he had spared no expense. A bronze plaque screwed into the stone announced to picnickers who might stray down between beers to the sacred river:

HERE BERTRAND CARRÉ GAVE HIS LIFE
THE FIRST TO RESIST IN TOURAINE
FALLEN FOR FRANCE
AT THE AGE OF FOURTEEN
MURDERED BY THE NAZIS
JUNE 21, 1940

With the exception of the name, the age, and the date, not a word of it was true.